Hands-On Data Science for Biologists Using Python

Hands-On Data Science for Biologists Using Python

Yasha Hasija and Rajkumar Chakraborty

CRC Press
Taylor & Francis Group
Boca Raton London New York

CRC Press is an imprint of the
Taylor & Francis Group, an **informa** business

First edition published 2021
by CRC Press
6000 Broken Sound Parkway NW, Suite 300, Boca Raton, FL 33487-2742

ISBN 13: 978-0-367-54679-3 (hbk)
ISBN 13: 978-0-367-54678-6 (pbk)

Library of Congress Cataloging-in-Publication Data
Names: Hasija, Yasha, author. | Chakraborty, Rajkumar, author.
Title: Hands on data science for biologists using Python / Yasha Hasija and Rajkumar Chakraborty.
Description: First edition. | Boca Raton : CRC Press, 2021. | Includes bibliographical references and index.
Identifiers: LCCN 2020044939 | ISBN 9780367546793 (hardback) | ISBN 9780367546786 (paperback) | ISBN 9781003090113 (ebook)
Subjects: LCSH: Biology–Data processing. | Python (Computer program language)
Classification: LCC QH324.2 .H373 2021 | DDC 570.285--dc23
LC record available at https://lccn.loc.gov/2020044939

Typeset in Times
by MPS Limited, Dehradun

Visit the Taylor & Francis Web site at
http://www.taylorandfrancis.com

and the CRC Press Web site at
http://www.crcpress.com

Contents

Preface

Yasha Hasija Ph.D.

Data science is rapidly becoming a vital discipline involving the use of big data to extract meaningful information. With the advent of high throughput technologies in the field of healthcare, it is becoming increasingly imperative for life science researchers to analyze the massive amount of data being generated. Researchers with little or no computational skills often find the task challenging. In order to overcome this challenge, we have meticulously drafted this book, using illustrative examples, as a stepwise guide to ease newcomers from the field of life sciences to the field of data science. We have chosen Python as our programming language of choice because of its easy accessibility on all operating systems, versatility, comprehensible interface, ease of use, object-oriented features, and wide range of applicability.

This book will serve as a beginner's guide for anyone interested in the basics of programming, data science, and Machine Learning. Every topic has an intuitive explanation of concepts and is accompanied by the implementation of the concepts using biological examples. This book can also serve as a handbook for biological data analysis using standard Python code templates for model building - facilitated with supplementary files for each chapter. The text is made to be as interactive as possible with accompanying Jupyter Notebooks for every section, to help readers practice the codes in their local systems. Each chapter is specially designed with examples.

The book is divided into two sections. The first section deals with an introduction to basic Python programming and a hands-on tutorial for data handling. Chapters in this section elaborate on the usage of some of the basic Python libraries and packages. One of the important libraries for life sciences data - Biopython - is explained in this section with examples of reading and writing various biological file formats, performing Pairwise and Multiple Sequence Alignments, handling protein and sequence data, etc. The subsequent sections elaborate on data handling using NumPy and Pandas, data visualization techniques, and dimensionality reduction methods that are common to all data analyzes and also provide illustrative examples for biological data.

Machine Learning is an integral part of several research projects today and has numerous applications in the present-day era. Almost all of the disciplines of technology have been transformed by Machine Learning and artificial networks, and life sciences are no exception, with Machine Learning applications in fields ranging from agriculture to diagnostics to personalized medicine to drug development to biological imaging - the list is mounting. The second section of the book deals with Python implementation in Machine Learning algorithms. Chapters in this section contain an introduction to Machine Learning to make readers comfortable with the various terminologies used in Machine Learning. This section also explores popular supervised and unsupervised Machine Learning algorithms - such as logistic regression, k-nearest neighbors, decision trees, random forests, support vector machines, artificial neural networks, convoluted neural networks, natural language processing, and k-means clustering - and shows their implementation in Python.

The book is written considering the need for biologists to learn programming in light of handling massive data, analyzing it, and deriving useful insights from it. I hope our readers will benefit from this hands-on book on data science for biologists using Python.

Yasha Hasija, Ph.D.

Author Bio

Dr. Yasha Hasija (B.Tech, M.Tech, Ph.D.) is an Associate Professor at the Department of Biotechnology and the Associate Dean of Alumni Affairs at the Delhi Technological University. Her research interests include genome informatics, genome annotation, microbial informatics, integration of genome-scale data for systems biology, and personalized genomics. Several of her works have been published in international journals of high repute, and she has made noteworthy contributions in the area of biotechnology and bioinformatics as author and editor of notable books. Her expertise, through her book chapters and conference papers, is of significance to other academic scholarship and teaching. She is also on the editorial boards of numerous international journals.

Dr. Hasija's work has brought her recognition and several prestigious awards - including the Human Gene Nomenclature Award at the Human Genome Meeting (2010) held in Montpellier, France. She is the project investigator for several research projects sponsored by the Government of India - including DST-SERB, CSIR-OSDD, and DBT. As Dr. Hasjia continues conducting research, her passion for finding the translational implications of her findings grows.

Mr. Rajkumar Chakraborty (B.Tech, M.Tech) received his Bachelor of Technology Degree in Biotechnology from the Bengal College of Engineering and Technology, West Bengal, India and completed his Masters of Technology Degree in Bioinformatics from the Delhi Technological University, Delhi, India. He is currently pursuing his Ph.D. in the field of bioinformatics. He was a part of the 4-member team which won "Promising Innovative Implementable Idea Award" at the SAMHAR-COVID19 Hackathon 2020 for innovating a solution towards drug repurposing against COVID-19. His research interests are in applied Machine Learning and the integration of big data in biological science.

1

Python: Introduction and Environment Setup

Why Learn Python

Before knowing about Python, we should first understand why people working in the area of life sciences should learn to program. As we are in the era of information technology, we have seen a massive explosion in biological data like sequences, annotations, interactions, biologically active compounds, etc. For instance, while this chapter was being written last April 2019, the Gene Bank (NCBI) - which is one of the largest databases for nucleotide sequences - contains 212 million sequences in its repository (https://www.ncbi.nlm.nih.gov/genbank/statistics/). EMBL, which is also a raw nucleotide sequence repository, contains 2,253.8 million annotated sequence data which are expected to double in about 19.9 months (https://www.ebi.ac.uk/ena/about/statistics). This extensive data is being generated by the advent of high-throughput technologies. For the analysis of this massive amount of data, we need the help of computers. Computers consist of a central processing unit (CPU), a primary memory, and a secondary memory storage device. The CPU is the component that does operations on the data stored in primary and secondary memory. Primary memory is as fast as the CPU and is designed to keep up with its speed, but it loses its memory as soon as the power is switched off. A secondary memory storage device can store data after the computer shuts down. These make up our digital assistant - which is pretty fast and accurate in its tasks and does not get bored with repetitive jobs. However, in order to assign the job to computers and to receive the desired output, we need to comprehend their language, which is also known as the programming language. Every biological research involves using different datasets and has unique problems to solve - from filtering, merging, subsetting, finding commonalities between lists, and may even require customization of data formats for preserving and using information. Programming gives a free hand to users to think and implement innovative algorithms and solve various problems.

Over time, data science has also found its applications in life sciences. Data science helps in finding patterns in a huge amount of structured or unstructured data which can help in providing valuable insights in almost all frontiers of biology - ranging from finding putative variations, predicting amino acid substitution consequences, diagnosing diseases quickly, predicting lead drug toxicity, predicting pharmacophores, personalized, or precision medicine, prediction in the field of protein secondary and tertiary structure, microRNA interaction with their targets, epigenetics, etc. The very first step in generating a hypothesis from a big amount of data is the curation of large datasets. A task like curating data is very tedious and time-consuming work. It consists of repetitive searching of data from certain database's websites, literature, and others. Here comes our digital assistant to the rescue, saving us from this tedious job as it can work much faster than how humans think and perform things manually. A 3.0-gigahertz CPU can process 3 billion instructions per second - that is an example of the tremendous power of computing.

The central theme of this book is to provide a practical approach to biologists in applying data science techniques on omics data. Data science usually consists of data analysis, data visualization, data preparation, Machine Learning, and more. We will discuss each aspect in relation to relevant biological problems along with their solutions - starting with basic Python programming so that readers can get accustomed to programming terminologies.

Programming skills are a valuable asset for any biologist. There are many programming languages that have been developed. Some are for instantaneous computation, website creation, and database generation, among others, and some are general-purpose programming languages that were developed to be used in a variety of application domains. Python is one example of a general-purpose programming language. Guido van Rossum developed it as a hobby in the Netherlands around 30 years ago and named it after a famous British comedian group called "Monty Python's Circus". Now, Python has applications in various domains like data science, web development, data visualization, and desktop applications, to name a few. Python is one of the popular programming languages in the data science and Machine Learning area, and it is community-driven. Since it has a very steady learning curve, it is recommended by many experts for beginners as their first programming language to learn. Primarily, Python has simple English-like readable syntax which is easily understandable by users. For example, if one wants to find the proportion of the amino acid Leucine with a symbol "L" contained in a protein sequence, the following Python code will do that:

```
Protein = "MKLFWLLFTIGFCWAQYSSNTQQGRTSIVHLFEWRWVDIALECERY"
Leu_contain = Protein.count('L')/len(Protein)
print(Leu_contain)
```

The code is very much similar to the English language. The first line is the protein sequence. The second line calculates the Leucine residues (denoted by the letter "L") by counting the number of times "L" appears in the sequence and then dividing it by the total length of the sequence. Moreover, at last printing the value, it turns out to be **0.108**

Thanks to the readability of Python codes, learners can concentrate on the concepts of programming and problems more than learning the syntax of the language. As Python is community-driven and it has one of the largest communities, Python has evolved to contain several important libraries that are pre-installed or are freely available to install. These libraries help in the quick and efficient development of complex applications, because these do not need to be written from scratch.

Another advantage of learning Python is that it can be used for various purposes due to the development of popular libraries, such as:

- Frameworks like Django, Flask, Pylons are used for creating static and dynamic websites.
- Libraries like Pandas, NumPy, and Matplotlib are accessible for data science and visualization.
- Scikit-Learn and TensorFlow are advanced libraries for Machine Learning and deep learning
- Desktop applications can be built using packages like PyQt, Gtk, and wxWidgets, among others.
- Modules like BeeWare or Kivy are taking the lead in mobile applications.

Learning programming is the same as learning a new language; we have to first understand the vocabulary and syntaxes. Next, we learn how to construct some meaningful but terse sentences. Using those sentences, we then form paragraphs, and finally, we write our own story. In this book, we will start with Python syntaxes and vocabulary. Then, we will construct small programs with biological relevance to help biologists learn programming with problems that are important to them.

Installing Python

We are using Python 3.7, which is the current and stable version of Python. Most of the operating systems either already have Python installed by default, or it can be downloaded from the Python Software Foundation's website (https://www.python.org/), where it is freely available. After installing Python, open the Python Shell in Windows or type "python3" in the terminal of Mac or Linux as follows:

```
Python 3.7.3 (v3.7.3:ef4ec6ed12, Mar 25 2019, 22:22:05) [MSC v.1916 64 bit
(AMD64)] on win32
Type "help", "copyright", "credits" or "license()" for more information.
»>Instructions are typed after "»>". Let us start typing our first instruction
and press enter.

»> print('Welcome to Python')
Welcome to Python
```

Our first instruction was simple - to print "Welcome to Python". If it runs correctly, then Python has been successfully installed and we are all set and ready to go!

Installing Anaconda Distribution

As we have discussed, Python has various packages that aid us in writing fewer lines of codes. Installing each package one by one is a time-consuming job. Moreover, because this book is centered on data science applications, we will require many widely used packages and along with their dependencies. For the sake of investing less time in setting up the coding environment, we will install the Anaconda distribution of Python. The Anaconda distribution comes with preinstalled packages for data science, and it is the most popular among data scientists. Most of the statistics, data visualization, and Machine Learning packages are built-in with the installation of Anaconda distribution. It is basically Python with a set of various useful tools and packages preinstalled within itself. We will also get IPython (i.e. an interactive Python shell) and Jupyter Notebook-like packages along with it. Jupyter Notebook will be used throughout this book for writing codes and executing these. Jupyter Notebook is a kind of interactive notebook based on IPython distribution. As a server-client application, the Jupyter Notebook App enables us to write, edit, and run our codes in notebooks through an internet browser. The application can be executed on a personal computer even without internet access. It comes with an Integrated Development Environment (IDE) which has autofill options for variables and packages. The Jupyter Notebook is also an easy way to share codes, so the codes used in this book may be downloaded and executed in the machines of users.

For more information about the Anaconda distribution, one can visit their official website (https://www.Anaconda.com/distribution/). To install Anaconda on the computer, go to (https://www. Anaconda.com/distribution/#download-section). Choose Python 3.x version, where x is equal to or greater than 7, and then download the graphical installer according to the user's operating system (i.e. Windows, Linux, or Mac OS). Follow the instructions for the graphical installer and keep all of the default options ticked.

Running the Jupyter Notebook

After installing the Anaconda distribution, we may now proceed to opening the Jupyter Notebook and then writing our first line of code. To do this, open the Anaconda command prompt in Windows or terminal for Linux or Mac OS users. Type "Jupyter Notebook", and the application should open on the default internet browser in this specific address: http://localhost:8888, only of course, if port 8888 is currently not in use. The user can also open Jupyter Notebook using the "Anaconda Navigator" by searching the same term in the applications. Please refer to the screenshot (Figure 1.1) below to demonstrate if we are on the same page or not.

The "Files" tab shows the browsable list of files and folder in the working directory. The "Running" tab displays the currently active Jupyter Notebook or terminals. The "Clusters" tab is for multiple assemblies of computers connected to a node. To create a new file, folder, or Jupyter Notebook, click the "New" button in the upper right corner of the page. Upon clicking the "New" button, a dropdown menu

FIGURE 1.1 A Screenshot of Root Directory Shown in a Jupyter Notebook.

FIGURE 1.2 New Notebook with Python 3.

will appear. Under the notebook section, choose "Python 3" to create a Python 3-compatible Jupyter Notebook (Figure 1.2).

As indicated here, the current name of the notebook is "untitled". To rename it, click on the "untitled" text itself. The cells will run Python 3 codes, as Python 3 was selected as the kernel. Try writing the same code that we have initially typed in the Python terminal:

```
print('Welcome to Python')
```

Click on the "Run" button above or press "Shift" + "Enter" to execute the code.

The following output should appear in the notebook:

```
Welcome to Python
```

In the event that the user has different cells in their Notebook, and the user runs the cells altogether, then the user can share their variables and imports among cells. This makes it simple to separate out the code into legitimate pieces without expecting to reimport libraries, reintroduce variables, or define functions in each cell.

The Jupyter Notebook has a few menus that the user can utilize to connect with their Notebook. The menus are as follows:

- File
  The File menu is used to create new notebooks, save notebooks, and open previously saved notebooks. Jupyter Notebook is typically saved in a ".ipynb" format, but the user can also save it in other formats by using the "Download as" option. Also, saving checkpoints options are automatically given.
- Edit
  The Edit menu consists of typical editing options like cut, copy, paste, merge cells, and others.
- View
  The View menu is useful in toggling the header and the toolbar.
- Insert
  The Insert menu is used for inserting cells below and above the current cell.
- Cell
  The Cell menu consists of running the cells and changing the type of cells.

- Kernel
  The Kernel option is mostly used in debugging to interrupt and to restart the Python 3 kernel.
- Widgets
  JavaScript widgets can be added to our cells to create dynamic content using Python. This menu is for saving and clearing the widget state.
- Help
  The Help menu is used for learning about Jupyter Notebook, its documentation, shortcuts, etc.

We can also add rich content in the Jupyter Notebook using markup language in the cells and change the cell type using the "Cell" menu to markdown. The markup language is a superset of HTML and is used for styling text, inserting maths equations, etc. To learn more about Jupyter Notebook, the user can always refer to the documentation.

The Building Blocks of Programs

In the next few chapters, we will learn more about vocabulary, syntax, and problem-solving with Python. We will discover the powerful abilities of Python and amalgamate those capabilities to create exciting programs.

There are some basic patterns used for building a program. These building blocks are not just for Python programs, rather, these are more or less the same for any programming language. These are discussed below:

Input: Accepting data/information from the user. Input might be from typing on the keyboard, reading from a file like Fasta Q or PDB, or even acquiring information from some form of a sensor - for example, from biomedical devices or color detection.

Output: Displaying the result, storing these in a file, or sometimes giving commands to other devices such as in robotics or automation.

Serial Execution: Executing statements one by one and according to the order that these are described in the script.

Conditional Execution: Checking for specific conditions, running, and/or skipping some of the commands.

Repeated Execution: Executing a group of statements continuously and maybe with slight differences.

Reuse: Writing a batch of instructions once, naming these, and then reusing those statements when needed throughout the program.

Errors in Python

Python gives rather detailed error messages by pinpointing the statement and library which are being used. Correcting or understanding errors are sometimes bothersome, but this process hones one into a successful programmer. There are different types of errors. Some are understandable by Python, and these can give alerts as warnings, but errors are not native to Python most of the time, so programs are sometimes executed with unexpected results. Here are three major types of errors in this discussion:

Syntax Error: These are the errors that are the most simple to understand and correct. These usually happen when the user scrambles the grammar rules of Python, and Python gets confused over these disarranged statements. Python will tell the user where the exact point of confusion is with the line and word and ask the user to correct this. For learners, this is the most common mistake or error message. Structuring the statements correctly is an essential requirement for proper execution.

Logic Errors: These are the errors in which Python does not understand the program and executes it with unexpected results. These happen when the user's statement is grammatically correct, but the meaning is not intentional. Logical errors are bugs, and the debugging process will help here. The user must look through all the steps to find the bug.

Semantic Errors: These types of mistakes happen when the user gives a grammatically correct statement in the proper order, but there is a problem in the program. For example, when the user tries to add or subtract a number with a string. This kind of operation is not possible and will raise a semantic error - pointing out the operation or statement.

Readers will encounter a lot of errors, and correcting these requires the skill of asking the questions discussed above.

With this prerequisite knowledge and environment setup, we are now ready to take deep dive into the exciting world of Python language and start writing our programs. In this book, the codes are explained in a step-by-step process so that these are understandable and applicable in solving individual problems. Data science techniques sometimes require a great depth of mathematical and statistical understanding. These are beyond the scope of this book. However, we will provide a necessary and intuitive explanation in every section. We hope that this wealth of knowledge helps the readers understand and appreciate the usability of programming for a biologist, the features of Python language, the combination of Python and data science for biologists, and ultimately, discover the fun way to learning all of this.

Exercise

>sp|Q9SE35|20-107

```
QSIADLAAANLSTEDSKSAQLISADSSDDASDSSVESVDAASSDVSGSSVESVDVSGSSL
ESVDVSGSSLESVDDSSEDSEEEELRIL
```

1. Why do biologists need to learn programming?
2. What are the types of errors generally encountered by programmers?
3. Comment on the building blocks of a typical computer program.
4. What are the advantages of installing Anaconda distribution over vanilla Python?
5. What are the applications of data science in the field of biology?
6. Write the functions of menus present in the Jupyter Notebook.
7. Calculate the serine (S) content in the given peptide sequence using the Python programming language.

2

Basic Python Programming

In this chapter, we will go through a basic understanding and overview of Python programming which is a prerequisite for any form of data analysis. Variables and operator, string, list and tuples, dictionary, conditions, loops, functions, and objects are some of the topics covered in this chapter.

Let us begin with a familiar syntax in Python which is used for commenting a statement. If a statement or a line begins with "#", then Python will ignore it. Comments are useful to make any code self-explanatory. We will use a lot of comments wherever required to make the code more understandable to readers.

Code:
```
#Let's print "Hi there!"
print('Hi there!')
```

Output:
```
Hi there!
```

After executing the above code in the Jupyter Notebook by pressing "ctrl + enter", the first statement or line starting with "#" will be ignored, and as a result, we will see "Hi there!". Therefore, the first line is a comment which describes that the code will print "Hi there!".

Datatypes and Operators

Datatypes

As the name suggests, datatypes are the different types of data, such as numbers, characters, and Booleans that can be stored and analyzed in Python. Among various datatypes, the four most common are:

- int (integers or whole numbers)
- float (decimal numbers or floating-points)
- bool (Boolean or True/False)
- str (string or a collection of characters like a text)

There are two important ways in which Python represents a number: int and float. Decimals numbers (float), such as 1.0, 3.14, −2.33, etc., will potentially consume more space than integers or whole numbers, like 1, 3, −4, 0, etc. Think of this way, if we take whole numbers between 0 and 1, then we will see only the two numbers 0 and 1, but, in the case of decimal numbers, we will get infinite numbers between 0.0 and 1.0. Next, we have a Boolean datatype, which is "True" or "False", and these are used in making conditions which we will learn shortly. Lastly, "str" or the string datatype is the datatype that biologists will need and encounter the most - whether it is the DNA, RNA, and/or protein sequences or names, most of them are text or strings. Therefore, we have a separate section for strings in this chapter. It is imperative to mention here that string-type data always remains inside quotes, i.e. ('<string data>'). For example, 'ATGAATGC' will be a string for Python.

To know the datatype of values, we can write "type(<value>)" to get its datatype in the Jupyter coding cell.

Code:
```
print(type(4)) # integer, or a whole number
print(type(4.0)) # floating point, or decimal number
print(type(True)) # boolean, or a True/False
print(type('ATGAATGC')) # means string, or 'a piece of text'
```

Output:
```
<class 'int'>
<class 'float'>
<class 'bool'>
<class 'str'>
```

In the statements, we can see that 4 is the "int" type, whereas 4.0 is in "float". For now, ignore the word "class". We will learn about this in the succeeding parts of this book. The key takeaways here are the datatypes and the method in identifying the datatype of value in Python.

Code:
```
# Addition
print(4 + 6) #Integer + Integer
print(4 + 6.0) #Integer + Float
# Subtraction
print(6-3) #Integer - Integer
print(6-3.0) #Integer - Float
# Multiplication
print(2 * 5) #Integer * Integer
print(2 * 5.0) #Integer * Float
# Division
print(24/3) #Integer / Integer

# Power
print(2**8) #Integer ** Integer
print(2**8.0) #Integer ** Float
# '%' or modulo operator, also known as the modulo or remainder operator gives
# the remainder of two numbers which are not a factor of each other.
8%3 #Integer % Integer
```

Output:
```
10
10.0
3
3.0
10
10.0
8.0
256
256.0
2
```

Operators

In this section, we will discuss some of the standard operators in Python. We are familiar with some of the operators like "+", "−", "*", "/", "=", and "**".

TABLE 2.1

Some Common Operators in Python.

Symbol	Name
+	Addition
−	Subtraction
*	Multiplication
/	Division
**	Power
%	Modulo
=	Equal to

Operations with an integer and float will always return float-type results, and operations with two integers will return integers, except for division where these will still return a float type. Subsequently, we can attain an integer-type for division by using an integral division operator (i.e."//").

Order of Operation – PEMDAS

For a complex calculation involving two or more operators, the order of operation is determined by the rule of PEMDAS:

1. Parentheses ()
2. Exponent **
3. Multiplication *
4. Division / // %
5. Addition +
6. Subtraction −

After PEMDAS, the order goes from left to right. For example, try to evaluate "2 + 5*4/2". According to "PEMDAS", first calculate "5*4", then "5*4/2", and lastly "2 + 5*4/2 = 12". Now, if the user has to break this order, they can use the Parentheses as used in pen and paper-solving of equations.

Variables

Variables in Python are like the variables of algebra in mathematics. We think of a variable as a box with a name on it that can hold any value or datatype. Variables can also inherit all the properties of the value stored inside it. Variables consist of two parts: the name and the value. We assign a name for the value by using an equal to "=" operator. The name is on the left side, and the value is on the right side.

Code:
```
length_of_gene = 1300
print (length_of_gene)
```

Output:
```
1300
```

Once we assign a variable, then we can recall them. In the example below, we can see that variables: "length_of_gene" and "length_of_introns" are assigned and then are used for finding the mRNA length and storing it in another variable called the "length_of_mRNA".

Code:
```
length_of_gene = 1300
length_of_introns = 350
length_of_mrna = length_of_gene - length_of_introns
print(length_of_mrna)
```

Output:
```
950
```

From this point forward, we will use these variables in other programs.

Variables make our programs clear enough to read, and these are reusable. For example, if the user has to use a long protein or nucleotide sequence, then it would not be wise to write it every time. Therefore we can assign it to a variable, and we can reuse this every time it is required. Variables can be assigned to other variables, reassigned anytime to different values, and also allocated to another variable. Let us explain this in code:

Code:
```
some_var = 100
another_var = some_var
some_var = 300
length_of_gene,length_of_introns = 1300,350
```

In the code, "another_var" is assigned the same as "some_var", and the next line "some_var" is reassigned to another value. When assigning a new value to the variable, the old value will be forgotten and, thus, cannot be retrieved. This reassigning of a variable can also be done with a non-identical datatype. For example, a variable containing an integer can be reassigned to a variable containing string and vice versa. This property is not true for many other programming languages. In the last statement, two variables are assigned values in the same statement - which is also one of the unique points of Python that sets it apart from any other programming language. Last but not the least, variable names are case sensitive - for example, a variable name "protein_id" cannot be called "Protein_ID" or "PROTEIN_ID".

Rules for Variable Naming

In Python, nomenclature can be assigned to a variable with a set of rules:

- Variable names must start with a letter or an underscore.
- The rest of the name should consist of letters, numbers, or underscores. No special characters like "@", ".", etc. are permissible.
- Python variables are case sensitive, as discussed earlier.
- 33 words are prohibited from being used as a variable name, because these are in Python 3.7's vocabulary and are known as keywords. All Python keywords are listed in Table 2.2.

TABLE 2.2

Keywords in Python

False	else	import	pass	Yield
None	break	except	in	Raise
True	class	finally	is	return
and	continue	for	lambda	try
as	def	from	nonlocal	while
assert	del	global	not	with
elif	if	or		

Most Python programmers prefer to name the variables with the following guidelines:

- Most variables should be in "snake_case", which means there is an underscore between words.
- Most variables are in lowercase other than constants.
- CamelCase is used for defining class or functions. Please note that we have a dedicated section for classes and functions in this chapter, so just remember this part for now.

Strings

For computer programmers, strings are the collection of characters or, more commonly, any texts. In bioinformatics studies, handling strings is very common - like sequencing files, finding patterns in the sequences, data-mining from texts, processing data from various file formats, etc. By enclosing a sequence of characters between a pair of single quotes, double quotes, triple-single quotes, or triple-double quotes, a string object can be constructed in Python. While characters enclosed between single or double quotes can only have a single line, characters between triple-single or triple-double quotes can have multiple lines. Let us take a look at the following example:

Code:

```python
# A string within a pair of single quotes
seq_1 = 'ATGCGTCA'
print(seq_1)
print('---------')


# A string within a pair of double quotes
seq_2 = "ATGCGTCA"
print(seq_2)
print('---------')


# A string within a pair of triple single quotes
seq_3 = '''ATGCGTCA'''
print(seq_3)
print('---------')


# A string within a pair of triple double quotes
seq_4 = """ATGCGTCA"""
print(seq_4)
print('---------')
```

```
# A string within a pair of triple single quotes, can have multiple lines
seq_5 = '''MALNSGSPPA
IGPYYENHGY'''
print(seq_5)
print('---------')

# A string within a pair of triple double quotes, can have multiple lines
seq_6 = """IGPYYENHGY
IGPYYENHGY"""
print(seq_6)
```

Output:
```
ATGCGTCA
---------
ATGCGTCA
---------
ATGCGTCA
---------
ATGCGTCA
---------
IGPYYENHGY
---------
IGPYYENHGY
IGPYYENHGY
```

The characters should be enclosed within the same type of quote - usually single or double quotes - for defining a string datatype.

Escape Sequence Characters

Supposing the user has to print text in different lines using double or single quotes, as programmers generally like to use double or single quotes for defining text. Perhaps, the user wants to use quotes inside quotes or more. Here, we have escape sequence characters to come to our aid. These are indeed not unique to Python and are found in various other languages. Given below is the list of escape sequences and their meanings:

Escape Sequence	Meaning
\newline	Ignored
\\	Backslash (\)
\'	Single quote (')
\"	Double quote (")
\a	ASCII Bell (BEL)
\b	ASCII Backspace (BS)
\f	ASCII Formfeed (FF)
\n	ASCII Linefeed (LF)
\r	ASCII Carriage Return (CR)
\t	ASCII Horizontal Tab (TAB)
\v	ASCII Vertical Tab (VT)
\ooo	ASCII character with octal value *ooo*
\xhh...	ASCII character with hex value *hh...*

Although most of these are not commonly used, we will try out some of the examples.

Code:
```
# Escape Sequence Characters
print(' Hey Ashok, "How\'re you?" ') #escaping single quotes
print('---------')
print('First line\nSecond line') #escaping new line
print('---------')
print('\\') #escaping Backslash
```

Output:
```
Hey Ashok, "How're you?"
---------
First line
Second line
---------
\
```

In this example, applications of escaping characters are shown. they are mostly used for writing text files using Python.

String Indexing:

A string is a collection or sequence of characters, so it is possible in Python to grab single characters as well as a part of the text by using their indexes. For grabbing the character, we have to place the index number inside the square bracket pair after the string name.

Given below is an example of String Indexing in the DNA sequence "ATGCGTCA" to print the second nucleotide.

It may be noted that the index of any string starts with 0, starting with the leftmost character - meaning that the index of first nucleotide "A" is 0, that of the second nucleotide "T" is 1, and so on. In backward indexing, the indexing starts with -1 from the rightmost character, meaning the backward index of last nucleotide "A" is -1, that of second last nucleotide "C" is -2, and so on.

Code:
```
dna_seq = 'ATGCGTCA'
print(dna_seq[1])
```

Output:
```
T
```

In the output we got "T", but the first nucleotide was "A". It is because, unlike our customary practice, Python counting starts with zero.

Another example of character index for Python is shown in the figure below, where the first row is the sequence; the second row is the forward index of nucleotides, and the third row shows the backward index (Figure 2.1):

A	T	G	C	G	T	C	A
0	1	2	3	4	5	6	7
-8	-7	-6	-5	-4	-3	-2	-1

FIGURE 2.1   String Indexing in Python.

Below is the code for extracting the first character of the string:

Code:
```
# Extracting the first nucleotide
dna_seq = 'ATGCGTCA'
print(dna_seq[0])
print('---------')

# Grab the second last nucleotide
print(dna_seq[-2])
```

Output:
```
A
---------
C
```

To grab a part of a text or string, the annotation used is "string_name[start: end]", where "start" is the starting index, the "end" is the index extending up to the provided number, but not including it.

- dna_seq[3:6] is "CGT" - characters starting at index 3 and extending up to but not including index 6
- dna_seq[3:] is "CGTCA" - leaving a blank for either index defaults to the start or end index of the string
- dna_seq[:] is "ATGCGTCA" - emptying both fields always produces a copy of the whole string
- dna_seq[1:5] is "TGCGTCA" - an index that is too big is truncated to string length
- dna_seq[:-4] is "ATGC" - selecting up to but not including the last four characters
- dna_seq [-4:] is "GTCA" - starting with the fourth character from the right end to the right end

String Concatenation

There are a few ways to concatenate or join strings. The easiest and most common way to add join strings by using the plus symbol (+) or, in simplest terms, by simply adding them.

Code:
```
#String concatenation
dna_1 = 'ATGCGTCA'
dna_2 = 'ACTGCGTC'
full_dna = dna_1 + dna_2
print('The sequence of DNA is 'full_dna)
```

Output:
```
The sequence of DNA is ATGCGTCAACTGCGTC.
```

We can add any number of strings using the "+" operator. An important thing to note here is that all of the datatypes should be strings while adding strings - for example, if we add a string with an integer, like "ACTGCGTC" + 4, then there will be an error message suggesting that "str" type and "int" cannot be added. To add a number, we have to convert the number to "str" type by using str(number) function. While we cannot add integer with strings, we can print the same string multiple times using the "*" operator with an "int" datatype. For example, "ACTGCGTC"*2 will double the string into " ACTGCGTCACTGCGTC".

Commands in Strings

Various commands are available to make the desired modifications in strings or to carry out analyses. We will discuss some of the most common methods in this section. Remember that these methods do not modify the string itself but, rather, produce a new string, because the string is an immutable datatype.

Let us return to the Jupyter Notebook and try out the following codes:

Code:
```
#Converting a string into lowercase letters
dna_seq = 'ATGCGTCA'
print(dna_seq.lower())
print('---------')
print(dna_seq)
```

Output:
```
atgcgtca
---------
ATGCGTCA
```

In the example above, the lower() method is used. It reverts the strings in lowercase letters. We can also observe that the original variable "dna_seq" is not changed after applying the lower() method on it. In the same way, using the command str.upper() will change the string into uppercase letters.

A few more commands for string alteration include count(), find(), and len(). Their usage is described below:

Code:
```
dna_seq = 'ATGCGTCA'
print(dna_seq.count('A')) #str.count()counts all the occurrences of the
selected string in the parent string.
print(dna_seq.find('GT')) #str.find() returns the index of the first occurrence
of the selected string in the parent string.
print(len(dna_seq)) # len()returns the length of the string.
```

Output:
```
2
4
8
```

In the above examples, "len()" is a function that returns the length of the string. There is a primary method called str.split() which is very frequently used for extracting data from delimited text file formats like CSV, TSV, etc. CSV stands for comma-separated-values, where values of each column are separated by a comma delimiter, and TSV stands for tab-separated-values, where the values of each column are separated by a tab delimiter

```
Pregnancies,Glucose,BloodPressure,SkinThickness,Insulin,BMI,DiabetesPedigreeFunction,Age,Outcome
6,148,72,35,0,33.6,0.627,50,1
1,85,66,29,0,26.6,0.351,31,0
8,183,64,0,0,23.3,0.672,32,1
1,89,66,23,94,28.1,0.167,21,0
0,137,40,35,168,43.1,2.288,33,1
```

FIGURE 2.2 Example of a CSV-Formatted File.

Figure 2.2 is an example of a CSV-formatted file, where the first row is known as the header row and consists of column names, and the rest of the rows are instances that have values separated by a comma for each column. We can extract the values of each row if we consider each row as a string using str.split() method:

Code:
```
#str.split()
first_row = '6,148,72,35,0,33.6,0.627,50,1'
Pregnancies,Glucose,BloodPressure,
SkinThickness,Insulin,BMI,DiabetesPedigreeFunction,
Age,Outcome = first_row.split(',')
print(Glucose)
print(BloodPressure)
print(Insulin)
print('---------')
print(first_row.split(','))
```

Output:
```
148
72
0
---------
['6', '148', '72', '35', '0', '33.6', '0.627', '50', '1']
```

Let us study the above code line by line. We assign the first observation (i.e. the second row of the CSV file shown in Figure 2.2) to a variable named "first_row". Second, we use the multiple-variables assigning feature of Python for setting up each column as variable and first observations as their values, respectively. Here the split(',') method collects a string and returns a list of values that are split by commas. We can print the variables named after the columns in the header of the CSV file. Also, the last line of the output is a list of split values. The list is a particular datatype in Python, which we are going to discuss in the next section. We are barely grazing the surface, and it should be noted that there are other exciting methods present for the string datatype. Readers can always refer to the Python documentation to find all of the methods available for strings.

Lists and Tuples

Now that we are familiar with the datatypes like integers, strings, and Booleans, we will discuss two more datatypes in this section - lists and tuples.

Lists and tuples store or hold multiple values of any datatype-like containers. These are also known as data structures, because they store data in a particularly convenient way so that these can be retrieved easily later.

Lists

While strings are a collection or sequence of characters, lists are a series of values that are more like arrays in other programming languages but are more comparatively flexible. Values in lists are known as items or elements. Some essential features of lists are:

- Lists are ordered. A list notes the order of the items inserted and can be retrieved later.
- Objects in a list can be accessed with an index.
- Lists can contain any entity - numbers, strings, tuples, and even other lists.
- Lists can be modified or mutable. Changes may be made in the list; new items can be added; existing items removed or revised.

There are different ways to build a new list. The best way is to put the elements in a square bracket ('['and']') separate them by commas.

Code:
```
# Creating an empty list called "list"
list = [ ]
# Adding the values inside the List
list = [ 1, 2, 3, 4, 5]
# Printing the List
list
```

Output:
```
[1, 2, 3, 4, 5]
```

Lists can hold any datatype or objects and can be assigned to any variable

Code:
```
# Adding the values irrespective of their datatype: Integer, String, float.
list = [1, 2, 3,'Metformin', 4.0, 4/2]
list
```

Output:
```
[1, 2, 3, 'Metformin', 4.0, 2.0]
```

Code:
```
# Creating a list called drug_name
drug_name = ['Metformin', 'Acarbose', 'Canagliflozin', 'Dapagliflozin']
print(drug_name)
```

Output:
```
['Metformin', 'Acarbose', 'Canagliflozin', 'Dapagliflozin']
```

Accessing Values in a List

Like strings, list items also have indexes starting with "0" for forward indexing and "-1" for backward indexing (Figure 2.3)

['Metformin'	'Acarbose'	'Canagliflozin'	'Dapagliflozin']
0	1	2	3
-4	-3	-2	-1

FIGURE 2.3 Python List Indexes.

We can access items inside a list using brackets [] and indexes.

Code:
```
# Accessing the elements in the list
print(drug_name[0]) # Metformin
```

```
print(drug_name[-2]) # Canagliflozin
# Calculating the length of the list.
print(len(drug_name))
# Calculating the length of individual elements in a list
print(len(drug_name[0])) # length of "Metformin" is 9
```

Output:
```
Metformin
Canagliflozin
4
9
```

Nested List

A list can hold any type of data, and a list can also hold lists on its own. A list inside another list is called a nested list. Nested lists are often used for storing and accessing two-dimensional matrix data. Figure 2.4 shows access to the nested list elements and their representation as two-dimensional data.

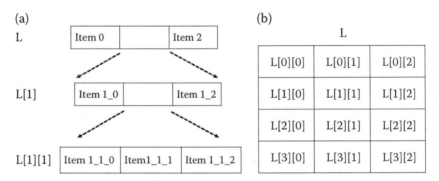

FIGURE 2.4   a) Nested List Indexes and b) the Representation of Nested Lists as Two-Dimensional Data.

Let us dive into Python to capture some elements from nested lists.

Code:
```
#list inside list be like list = [[],[],[]]
list_in_list = [['list0_0','list0_1','list0_2'],
['list1_0','list1_1','list1_2']]
#grab 'list1_2'
print(list_in_list[1][2])
```

Output:
```
list1_2
```

Accessing subparts of a list is known as slicing the list. A strings slice operator is also used for slicing lists. Similar to strings, if the first index is omitted, then the slice starts at the beginning; the same thing applies to lists as well. If the second one is omitted, then the slice will proceed to the end. Therefore, if both are omitted, then the slice becomes a copy of the entire list.

Code:
```
drug_name = ['Metformin', 'Acarbose', 'Canagliflozin', 'Dapagliflozin']
# first index is inclusive (before the :) and last (after the :) is not.
# not including index 2
print(drug_name[0:2])
# everything up to index 3
print(drug_name[:3])
# index 1 to end of list
print(drug_name[1:])
# Coping whole list
print(drug_name[:])
```

Output:
```
['Metformin', 'Acarbose']
['Metformin', 'Acarbose', 'Canagliflozin']
['Acarbose', 'Canagliflozin', 'Dapagliflozin']
['Metformin', 'Acarbose', 'Canagliflozin', 'Dapagliflozin']
```

We can use the "+" operator to concatenate two lists and the "*" operator to repeat a list for any specified number of times.

Code:
```
diabetes_drug = ['Metformin', 'Acarbose', 'Canagliflozin', 'Dapagliflozin']
pcos_drug = ['Drospirenone','Metformin', 'Letrozole', 'Norgestimate']
# Concatenating Lists using '+' operator
all_drugs = diabetes_drug + pcos_drug
print(all_drugs)
print('---------')
# Use of '*' operator in list
drugs = ['Metformin']*4
print(drugs)
```

Output:
```
['Metformin', 'Acarbose', 'Canagliflozin', 'Dapagliflozin', 'Drospirenone',
'Metformin', 'Letrozole', 'Norgestimate']
---------
['Metformin', 'Metformin', 'Metformin', 'Metformin']
```

Methods with Lists

Python provides some built-in methods for lists, such as:

- count() methods will produce the total number of occurrences of an item in the list.
- index() will provide the index of an item.
- append() adds an item at the end of the list.
- remove() will remove the first occurrence of the item in the list.
- pop() will remove the item at the index provided by the user.
- min(), max(), and sum() will provide the minimum, maximum, and the sum of the lists constituting number values.
- len() will provide the total number of items in the list.

We will use the list "all_drugs" from the previous example in the examples below:

EXAMPLE 1:
```
all_drugs = ['Metformin', 'Acarbose', 'Canagliflozin', 'Dapagliflozin', 'Drospirenone', 'Metformin', 'Letrozole', 'Norgestimate']:
```

Code
```
print(all_drugs.count('Metformin'))
print(all_drugs.index('Canagliflozin'))
all_drugs.append('New_Drug')
print(all_drugs)
all_drugs.remove('Metformin')
print(all_drugs)
# Remove item at the index
# this function will also return the item you removed from the list
# Default is the last index
print(all_drugs.pop(0))
print(all_drugs)
print(all_drugs.pop(0))
print(all_drugs):
```

Output
```
2
2
['Norgestimate', 'Metformin', 'Metformin', 'Letrozole', 'Drospirenone', 'Dapagliflozin', 'Canagliflozin', 'Acarbose', 'New_Drug']

['Norgestimate', 'Metformin', 'Letrozole', 'Drospirenone', 'Dapagliflozin', 'Canagliflozin', 'Acarbose', 'New_Drug']

Norgestimate

['Metformin', 'Letrozole', 'Drospirenone', 'Dapagliflozin', 'Canagliflozin', 'Acarbose', 'New_Drug']
```

EXAMPLE 2:

Code:
```
# for list containing numbers only
number_list = [3, 7, 2, 11, 8, 10, 4]
print(min(number_list), max(number_list), len(number_list), sum(number_list)):
```

Output:
```
2 11 7 45
```

Sorting Lists

List.sort() method can be used to sort a list of numerical values in increasing or decreasing order or a list of string in A-Z or Z-A alphabetical order.

Code:
```
number_list = [3, 7, 2, 11, 8, 10, 4]
# Sorting and altering original list
# increasing order
number_list.sort()
print(number_list)
# Sorting and Altering original list
# decreasing order
number_list.sort(reverse = True)
print(number_list)

# Sorting and Altering original List
# A-Z
all_drugs.sort()
print(all_drugs)

# Sorting and Altering original List
# Z-A
all_drugs.sort(reverse = True)
print(all_drugs)
```

Output:
```
[2, 3, 4, 7, 8, 10, 11] [11, 10, 8, 7, 4, 3, 2]

['Acarbose', 'Canagliflozin', 'Dapagliflozin', 'Drospirenone', 'Letrozole',
'Metformin', 'Metformin', 'Norgestimate']

['Norgestimate', 'Metformin', 'Metformin', 'Letrozole', 'Drospirenone',
'Dapagliflozin', 'Canagliflozin', 'Acarbose']
```

Difference Between Lists and Tuples

Lists and tuples are both collections or sequences of any type of values or objects which are stored in a specific order. Lists and tuples, in most cases, are identical with a few variations; therefore, in this section, we will understand tuples in reference to their differences with lists.

- Lists and tuples have different syntaxes. Lists are enclosed by brackets [], whereas tuples are enclosed by parenthesis ().

Code:
```
list_of_numbers = [1,2,3,4]
tuple_of_numbers = (1,2,3,4)

print(list_of_numbers)
print(tuple_of_numbers)
print(type(list_of_numbers))
print(type(tuple_of_numbers))
```

Output:
```
[1, 2, 3, 4]
(1, 2, 3, 4)
```

```
<class 'list'>
<class 'tuple'>
```

- Lists are mutable, whereas tuples are immutable. We can change or modify the items in a list, but we cannot change or modify the values stored in tuples. Python will report an error if we try to modify any tuple.

list_of_numbers = [1,2,3,4]

Code:
```
# Mutable List vs Immutable Tuple
list_of_numbers[2] = 5
print(list_of_numbers)

print('-----------------------------------------------------------------')
tuple_of_numbers = (1,2,3,4)
tuple_of_numbers[2] = 5
```

Output:
```
[1, 2, 5, 4]
-----------------------------------------------------------------
TypeError                                  Traceback (most recent call last)
<ipython-input-175-c46bde4c191e> in <module>
      3 print(list_num)
      4
----> 5 tuple_of_numbers[2] = 5

TypeError: 'tuple' object does not support item assignment
```

In the example here, the error clearly explains that tuple objects cannot be mutated.

- Tuples use less memory than lists.
- Python has more built-in methods for lists than tuples.

At a first glance, it may appear that lists can always substitute tuples. However, tuples are data structures that are incredibly useful. The choice of using a tuple over a list implies that the data should not be modified.

Dictionary in Python

Dictionaries are a type of data structure implemented within Python and are commonly known as hash tables, or hashmaps, in other programming languages. A dictionary is a collection of "key: value" pairs, where a single value is mapped with a key. There are different ways to build a dictionary. The best way is to put the "key: value" pairs inside curly brackets {}. An empty dictionary can be declared by these curly brackets {} only.

Code:
```
protein = {}
protein = {'Uniprot_ID':'P22103','Name':'Pneumadin','seq':'AGEPKLDAGV',
'seq_length': '10'}
```

```
print(protein)
print(type(protein))
```

Output:
```
{'Uniprot_ID': 'P22103', 'Name': 'Pneumadin', 'seq': 'AGEPKLDAGV', 'seq_
length': '10'}
<class 'dict'>
```

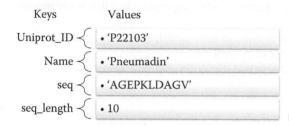

FIGURE 2.5   Python Dictionary Key: Value Pairs.

The essential properties of the Python dictionary are as follows (Figure 2.5):

- Dictionaries are unordered. Key: value pairs contained in the dictionary are not kept in any specific order.
- Dictionaries can be updated and are changeable or mutable. They can be modified, expanded, and reduced as required.
- The values of dictionaries are accessed or recorded by the keys. As "key: value" pairs are not stored in an ordered way, indexes cannot be accessed.
- One key can only appear once in a dictionary. The keys are, therefore, always unique in the dictionary.
- Keys of dictionaries should contain immutable datatypes - such as strings, numbers, tuples, etc.
- The values in a "key: value" pair of a Python dictionary can be of any datatype - such as numbers, strings, lists, or even dictionaries. A dictionary inside a dictionary is also known as a nested dictionary.

Stored values in a dictionary can be obtained using the key in square bracket. Let us study the example below:

protein = {'Uniprot_ID':'P22103','Name':'Pneumadin','seq':'AGEPKLDAGV','seq_length': '10'}

Code:
```
# Accessing the elements in a dictionary
print(protein['Name']) # Pneumadin
print(protein['seq']) # AGEPKLDAGV
print(protein.get('Name'))
print(protein['structure']) # KeyError
```

Output:
```
Pneumadin
AGEPKLDAGV
Pneumadin
```

```
KeyError                                Traceback (most recent call last)
<ipython-input-93-0623c5b5d512> in <module>
      3 print(protein['seq']) # AGEPKLDAGV
      4 print(protein.get('Name'))
----> 5 print(protein['structure']) # KeyError

KeyError: 'structure'
```

In the example, we can observe that Python gave a KeyError when we input a key that is not present in the dictionary.

A dictionary may contain another dictionary, which may, in turn, contain other dictionaries, and so on. This configuration of dictionaries is called nested dictionaries.

In the next example, we will learn how to create a nested dictionary and how to access its values.

Code:
```
# syntax dict = {key1:{},key2:{}}
proteins = {'Protein1':{'Uniprot_ID':'P22103','Name':'Pneumadin'},
            'Protein2':{'Uniprot_ID':'P02729','Name':'Urine
            glycopeptide'}}

#Grab protein2
print(proteins['Protein2'])
print('---------')
# Grab Protein2 Name
print(proteins['Protein2']['Name'])
```

Output:
```
{'Uniprot_ID': 'P02729', 'Name': 'Urine glycopeptide'}
---------
Urine glycopeptide
```

For fetching values in a nested dictionary, we have to refer keys of every level towards the value in square brackets, like dict[level_1_key][level_2_key], and so on.

Dictionary values can be changed by accessing the value to be changed by its key and by assigning a new value to it.

Code:
```
#Add or Update Dictionary Items
protein = {'Uniprot_ID':'P22103','Name':'Pneumadin','seq':'AGEPKLDAGV',
#x00027;seq_length':10}
print('Before: ',protein)
protein['Name']='New_name'
print('After: ',protein)
```

Output:
```
Before: {'Uniprot_ID': 'P22103', 'Name': 'Pneumadin', 'seq': 'AGEPKLDAGV',
'seq_length': 10}
After: {'Uniprot_ID': 'P22103', 'Name': 'New_name', 'seq': 'AGEPKLDAGV',
'seq_length': 10}
```

Elements can be removed from a dictionary using "del" keyword of Python and the pop() method, which is a built-in method of the Python dictionary. We can empty a dictionary using the clear() method.

Code:

```
protein = {'Uniprot_ID':'P22103','Name':'Pneumadin','seq':'AGEPKLDAGV',
'seq_length':10}

# Deleting an element
del protein['Uniprot_ID']
print(protein)
print('---------')
# Popping an item
protein.pop('seq_length')
print(protein)
print('---------')

# Clearing the entire dictionary
protein.clear()
print(protein)
```

Output:

```
{'Name': 'Pneumadin', 'seq': 'AGEPKLDAGV', 'seq_length': 10}
---------
{'Name': 'Pneumadin', 'seq': 'AGEPKLDAGV'}
---------
{}
```

A dictionary can be copied to another dictionary using the copy() method.

Code:

```
#copying a dictionary
protein = {'Uniprot_ID':'P22103','Name':'Pneumadin','seq':'AGEPKLDAGV',
'seq_length':10}
same_protein = protein.copy() # Coping to new variable
print(protein)
print('---------')
print(same_protein)
```

Output:

```
{'Uniprot_ID': 'P22103', 'Name': 'Pneumadin', 'seq': 'AGEPKLDAGV', 'seq_-
length': 10}
---------
{'Uniprot_ID': 'P22103', 'Name': 'Pneumadin', 'seq': 'AGEPKLDAGV', 'seq_-
length': 10}
```

The lists of all of the keys and values of a dictionary can be obtained using the keys() and values() method.

Code:

```
# Dictionary Keys and values
print(protein.keys()) #keys
print('-----------')
```

```
print(protein.values()) #values
```

Output:
```
dict_keys(['Uniprot_ID', 'Name', 'seq', 'seq_length'])
-----------
dict_values(['P22103', 'Pneumadin', 'AGEPKLDAGV', 10])
```

Conditional Statements

Until this point, we have recognized our programs as straightforward, but these did not seem to be smart, nor did these make any decisions. If we want a program to make decisions based on some conditions, then it will require conditional statements. Computers have only two conditions - "True" or "False" - similar to a light switch that has two conditions - "On" or "Off". Therefore, these "True" and "False" conditions constitute a particular datatype in Python referred to as Booleans.

Defining a condition always requires some kind of comparison - such as greater than, less than, or equal to. Some of the comparisons with their operators in Python are given below:

- Equal: a = = b
- Not Equal: a!= b
- Less than: a < b
- Less than or equal to: a < = b
- Greater than: a > b
- Greater than or equal to: a > = b

All of these comparisons result in the Boolean values "True" or "False".

Code:
```
#Booleans expressions and comparisons
# Comparing integers
print(100 = =100)
print(100 = =101)
print(200 > 100)
print(200 < 100)
# Comparing strings
print('hello'=='hello')
print('hello'=='hey')
print('hello'!='hey')
```

Output:
```
True
False
True
False
True
False
True
```

We can compare any datatype in Python - including comparisons between integers and strings - as illustrated in the example above. In the output, we have "True" and "False". These are not a string of characters; rather, they are predefined keywords of Python and are of the bool type.

Code:
```
# True and False are not strings, they are bools
print(type(100 = =100))
print(type(100 = =101))
print(type(True))
print(type(False))
```

Output:
```
<class 'bool'>
<class 'bool'>
<class 'bool'>
<class 'bool'>
```

Logical Operators

When we have to compare multiple conditions, we have to use a logical operator. Python has three logical operators - "and", "or", and "not" - their meanings in the Python program are the same as their definitions in English. Usually, logical operators operate on the following conditions:

Code:
```
#and, or, not
print(5 > 4 and 4 > 3)
print(True and False)
print(True and True)
print(False or True)
print(not True)
```

Output:
```
True
False
True
True
False
```

The "and" operator will only present true if both of the conditions are true. The "or" operator will show true if either of the conditions is true. Lastly, the "not" operator will simply give the opposite condition (i.e. it will produce false for true and true for false).

If and Else Statements

Often, we require to execute some statements only if some conditions are true. In this case, we have to use "if" and "else" statements.

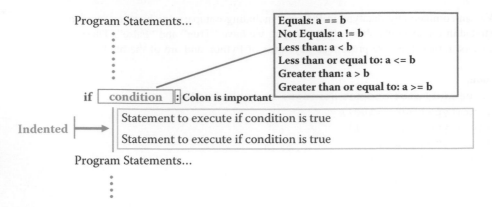

FIGURE 2.6   Syntax of an "If" Statement.

1. The "if" statement is used to execute a block of code if the specified condition is true.
2. The "else" statement is used with an "if" statement to execute a block of code if the specified condition is false.

Blocks are defined in Python by indentations after a colon in the statement above it. Indentation is a strict syntax of Python for building a block of the statement, which will often be encountered here. Figure 2.6 is a diagrammatic representation of an "if" statement with a condition and then a colon followed by the indented block of statements to be executed if the condition is true.

Let us study an example of comparing the expression of a gene in a controlled and treated environment below:

Code:
```
#if condition:
#     Statements to be executed if the condition is true

control_expression = 2.6
treated_expression = 3.5
if control_expression < treated_expression:
    print('Gene is upregulated')
```

Output:
```
Gene is upregulated
```

Now, we shall learn about an example comparing the expression of a gene in a controlled and treated environment with "If" and "else" statements.

Code:
```
#if (condition):
# Executes this block if condition is true
#else:
# Executes this block if condition is false

control_expression = 2.6
treated_expression = 3.5
if control_expression > treated_expression:
```

```
    print('Gene is downregulated')
else:
    print('Gene is upregulated')
```

Output:
```
Gene is upregulated
```

We can also insert "if" conditions within a block of an "if" statement for a problem involving multiple condition testing. This is also known as the "nested if" construct.

Let us compare the length of three genes in the following example:

Code:
```
#if (condition1):
    # Executes when condition1 is true
# if (condition2):
# Executes when condition2 is true
# if Block is end here
# if Block is end here
gene1_length = 2.6
gene2_length = 3.5
gene3_length = 2.8

if gene1_length < gene2_length:
print('Gene2 is longer than Gene1')
if gene3_length <gene2_length:
print('Gene2 is longest of the 3 Genes')
```

Output:
```
Gene2 is longer than Gene1
Gene2 is longest among the 3 Genes
```
We can check the presence of an item in lists or tuples by using the "in" keyword of Python.

Code:
```
#Check if an item is present in a Sequence
diabetes_drug = ['Metformin', 'Acarbose', 'Canagliflozin','Dapagliflozin']
if 'Metformin' in diabetes_drug:
print('Metformin is used for treating diabetes')
else:
print('Metformin is not use for treating diabetes')
```

Output:
```
Metformin is used for treating diabetes
```
The "in" keyword can also be used for checking the presence of a substring in a string.

Loops in Python

Computers are frequently used to automate repetitive tasks. Since iteration is frequent in programming, Python has several options to make this easier. The while loop is one form of iteration in Python.

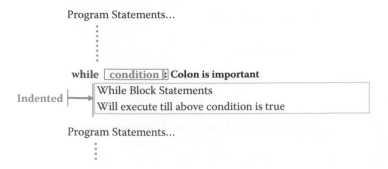

FIGURE 2.7   Syntax of a python while loop.

While loop

The syntax of a while loop is shown in the figure.

Loops in Python have a block of statements (recall how Python blocks are defined) which will be repeatedly executed until this meets a stop condition. In the syntax of the while loop, we see a condition which is called a stop condition. The modus operandi of a "while" loop is that the block of statements under the "while" statement will be executed until the stop condition is encountered.

A simple example of a "while" loop is printing numbers from 0 to 5, as illustrated below:

Code:
```
# Simple while loop
a = 0
while a < 6:
print(a)
a = a + 1
```

Output:
```
0
1
2
3
4
5
```

In the snippet of code above, we can observe that a variable "a" is initiated with value 0. Second, we have a "while" statement where the stop condition is "a" less than 6. The block under the "while" statement contains two statements - one for printing "a" and the other adds 1 increment to "a". When the block is first executed, the value of "a" is printed, and the same is increased from 0 to 1. Before rerunning the block, the stop condition of the while statement is checked. As "a" = 1 is still less than 6, the block will continue to run again. This cycle of repetitive execution is called a loop execution or iteration, and it stops when "a" will be equal to 6, and, hence, the condition of the loop fails. If we do not establish the stop condition or if we set in a condition that is never going to be false, then the loop becomes an infinite loop and will be executed countless times. To stop this, we have to restart the Python kernel in the Jupyter Notebook. If the asterisk on the left side of the cell persists for a long time, then this may imply that we may have entered an infinite loop. Let us see another example of the "while" loop for finding the longest DNA sequence below:

Code:
```python
# While loop
dna_list = ['AGGGC','ATTGGCCTT','AGGTTCC','GGCCTCA','TTTCCGGCTA','CCGCGTA']
#initiating variable to store longest length
longest = 0
#variable to store index of List
index=0

total_data=len(dna_list)

while (index < total_data):
    current=len(dna_list[index]) #lenght of current DNA

    if current > longest: #comparing the length of DNA with stored longest DNA
        longest=current # Storing the length of longest DNA
        longest_dna = dna_list[index] # Storing the longest DNA
    index=index+1

print(longest,longest_dna)
```

Output:
```
10 TTTCCGGCTA
```

In the example above, the comments explain each statement. As described above, the stop condition is the last index in the list. The last index of the list in the forward direction is 5, and the length of the list is 6, because it contains that number of elements. Thus, the stop condition is when the last index is increased by 1 and is equal to the length of the "total data" list or when both will be equal to 6.

For Loop

Loops are indefinite loops, because they run until a particular condition is false. However, we often prefer to loop through words in texts or items in lists or through keys and values of a dictionary. In such cases, we require definite loops like the "for" loop for accomplishing these tasks of iterating over sequence datatypes. The "for" loop runs through each item in a set of items. The syntax of a "for" loop is similar to the "while" loop: there is a "for" statement and a block of code under that statement.

Code:
```python
drug_name = ['Metformin', 'Acarbose', 'Canagliflozin','Dapagliflozin']
for drug in drug_name:
    print(drug)
```

Output:
```
Metformin
Acarbose
Canagliflozin
Dapagliflozin
```

The "for" loop iterates over each item in the list and runs the block statements for all of the items on the list. In simple English, this for loop can be translated as "run the block of code under the "for" loop for every drug in the drug list". In the "for" statement, the drug is the variable that changes with every iteration. This variable can have any name like other variables.

Code:
```
drug_name = ['Metformin', 'Acarbose', 'Canagliflozin','Dapagliflozin']
for temp in drug_name:
    print(temp)
```

Output:
```
Metformin
Acarbose
Canagliflozin
Dapagliflozin
```

Examine how the output with the variable name "drug" is the same as with the variable name "temp".

Breaking a Loop

In order to exit a loop before its full execution, a Python "break" statement is used. With a "break" statement, the program departs from the loop immediately, and the rest of the statements outside the loop block or body are executed.

Code:
```
# Braking a loop before its end
drug_name = ['Metformin', 'Acarbose', 'Canagliflozin','Dapagliflozin']
for temp in drug_name:
    print(temp)
    if temp = = 'Canagliflozin':
        break
```

Output:
```
Metformin
Acarbose
Canagliflozin
```

As the "if" condition is satisfied when "temp" variable is equal to "Canagliflozin", the execution of the "break" statement blocks the printing of "Dapagliflozin".

Range in Python

We can use a Python built-in function range() to execute a series of statements for a given number of times. Let us consider the application of the range() function for loops before we talk of "functions" in detail in the subsequent section. The "range" function produces a series from 0 to the number before the number in the "range".

Code:
```
# range function
for i in range(5):
    print(i)
```

Output:
```
0
1
2
3
4
```

The range() function offers a convenient way to repeat or iterate some tasks for a certain number of times.

Code:
```
for i in range(3):
    print('I love biology')
```

Output:
```
I love biology
I love biology
I love biology
```

By default, the range begins from 0. The range function can also start from any number, and this is done by defining a starting parameter at a specific number.

Code:
```
for i in range(6,10):
    print(i)
```

Output:
```
6
7
8
9
```

Functions

In programming, a sequence of statements that do particular work can be clubbed into a function. Functions are useful because these do not repeat the same statements in doing a particular task. Functions are a block of statements that perform a task and have a name so that we can identify them later. A function can also take arguments and returns some values. Let us define a simple function below:

Code:
```
def BioLove():
    print('I love Biology..!')
BioLove()
```

Output:
```
I love Biology..!
```

This is a simple function that just prints a few things. The "def" keyword is used to tell Python that we are going to define a function. After "def" keyword, we provide a name to the function with parenthesis;

here the given name is "BioLove()". Next, we put a colon, and by now we know what colon defines - a block of indented statements. The code block will run every time the function is called. The function is called by its name - in this case, like BioLove(). It will run the block of code under the "def" statement. Here it just prints "I love Biology...!".

We can pass values to a function. These values are known as the arguments of a function. Arguments are defined and passed inside the parenthesis of a function name. These arguments can be accessed inside the functions like variables.

Code:
```
#Functions with arguments
#Pass one arguments
def Drug(compound):
    print(compound,'is a Drug')

Drug('Metformin')
```

Output:
```
Metformin is a Drug
```

While defining a function, we have to choose the name of the argument. Arguments can be of any value or of any datatype. There, an argument "compound" is defined with the function, and this function prints the "compound is a Drug". Later, we can call this function by any drug name, and it will do the rest for printing it. Likewise, functions can take any number of arguments.

Code:
```
#Pass two arguments
def DrugDisease(Drug,Disease):
    print(Drug,'is used for treatment of',Disease)

DrugDisease('Metformin','type 2 diabetes')
```

Output:
```
Metformin is used for the treatment of type 2 diabetes
```

Types of Arguments

Unlike other languages, Python treats functional arguments in a rather versatile way. Various types of arguments can be assigned while defining a function. Type of arguments include:

- **Positional Arguments**
  Positional arguments are the most popular; we have also studied it in the example. Here, while defining a function and arguments, the positions are reserved for arguments. Later, when the function is called, the argument values should be placed in their respective positions. The drawback of positional arguments is that arguments must be transmitted in the order in which they were described.
- **Keyword Arguments**

In order to avoid ambiguity due to position, we can always determine arguments with their names. Thus, the order of arguments, in this case, does not matter any longer, because arguments are matched by name and not by position.

Code:
```
#Keyword Arguments
def DrugDisease(Drug,Disease):
    print(Drug,'is used for treatment of',Disease)
DrugDisease(Drug = 'Metformin',Disease = 'type 2 diabetes')
DrugDisease(Disease = 'type 2 diabetes',Drug = 'Metformin')
```

Output:
```
Metformin is used for the treatment of type 2 diabetes
Metformin is used for the treatment of type 2 diabetes
```

- **Default Arguments**

While defining a function, we can set default values for arguments. When the function is called without a relevant argument, then the default value will be used. In short, we can make the definition of certain arguments optional with default arguments.

Code:
```
def DrugDisease(Drug,Disease = 'type 2 diabetes'):
    print(Drug,'is used for treatment of',Disease)

DrugDisease('Metformin','PCOS')
DrugDisease('Metformin')
```

Output:
```
Metformin is used for the treatment of PCOS
Metformin is used for the treatment of type 2 diabetes
```

Returning Values

Functions can return values, by using the "return" keyword.

Code:
```
#Return Value

def MeanExpression(expressions):
    total = sum(expressions)
    mean = total/len(expressions)
    return mean

control = [2.6,3.0,2.5]
treated = [4.2,3.8,4.5]
control_mean = MeanExpression(control)
treated_mean = MeanExpression(treated)
if control_mean > treated_mean:
    print('Control gene in upregulated',control_mean-treated_mean)
else:
    print('Treated gene is upregulated',treated_mean-control_mean)
```

Output:
```
Treated gene is upregulated 1.4666666666666672
```

The function returns the mean expression for a list consisting of expression values of a gene with triplicate experiments. Mean expression of controlled and treated conditions are calculated. Next, these are compared to see whether the treated genes are upregulated or downregulated, and the difference in the means of the two is determined.

Once the "return" keyword is executed, the function terminates immediately and returns the value. Python functions can also return more than one value. Refer to the Python documentation for more details.

Python Built-In Functions and Modules

Python has many built-in functions. At this point, we have seen some - such as print(), type(), len(), and min(), among others.

Code:
```
print(type(164) #Returns the object type
print(max([1,2,3,4,5,6])) # Returns the max value
print(min([1,2,3,4,5,6])) # Returns the min value
print(sum([1,2,3,4,5,6])) # Returns the summation value
print(len('Data Science')) # Returns the length
print(len([1,2,3,4,5,6])) # Returns the length
```

Output:
```
<class 'int'>
6
1
21
12
6
```

Importing Modules

A Python module is a .PY extension script. Python modules are very easy to create. Simply build a file with a valid Python code and assign the '.py' extension with a name to the file. The modules in Python facilitate the modularization of code. Modules can be used while building Python programs by calling them using the "import" keyword. Using modules will make our programs more stable and efficient.

Python packages and libraries can be found for almost any programming task - whether this is for web development, database creation, image analysis, data science, statistics, Machine Learning, and the list goes on. Packages are a collection of modules, whereas libraries are a collection of functions. The Python Package Index (PyPI) is a repository for Python packages, and there are over 227,607 packages currently in the PyPI to make life easier for developers. We can install any of the packages available in the PyPI by using the "pip" installer. In an Anaconda distribution of Python, we can use both the "conda" and "pip" installer. For installing a package, we have to open the terminal or command prompt and type "pip install PackageName", or "conda install PackageName" in Anaconda distribution. The Anaconda distribution of Python has various preinstalled packages for data science applications. Once the packages are installed, we can use the modules using the "from" and "import" keywords. Let us generate some random numbers using the built-in "random" module of Python, as illustrated below:

Code:
```
import random
for i in range(10):
    x = random.random()
    print(x)
```

Output:
```
0.5634896957457116
0.9591754239667578
0.16992230877805425
0.7009904830629191
0.2580808526266217
0.8738009335474071
0.20978267149403984
0.4099941803239372
0.1990991894270332
0.43875165546687034
```

Computer-generated random numbers are called pseudorandom numbers. Pseudorandom numbers are not necessarily random, because mathematical calculations create them. However, it is impossible to distinguish them as incidental simply by looking at the numbers. The random module in Python is used to generate randomness. The module has the function random() which, by default, generates a random number between 0.0 and 1.0. We can change this range by passing two numbers as arguments in the random.random() function. This random module has various other functions. Let us examine some below:

Code:
```
print(random.randint(10, 20))
print(random.randint(10, 20))
drug_name = ['Metformin', 'Acarbose', 'Canagliflozin','Dapagliflozin']
print(random.choice(drug_name))
```

Output:
```
16
18
Canagliflozin
```

The function randint() from the random module generates a random integer between a specific range of integers. The choice() function, every time it is executed, chooses an item from the list randomly. Throughout this book, we will use various packages to make the readers' learning experience more comfortable.

Classes and Objects

Procedural programming ensures that a program is split into a variety of functions. Data is kept in a set of variables, and the functions manage them. When the programs expand, we result in a bunch of functions in place. We often have to copy and paste the lines of codes over and over. Dependability of one function on other functions can break our code while modifying a core function. Object-oriented programming can also solve this problem.

Through object-oriented programming or OOP, a set of associated variables and functions may be combined into a unit called an object. The object is made up of functions and variables. The functions present in an object are called methods. Two critical elements of OOP are classes and objects. A class is a template model for the creation of individual objects. For example, we can have an mRNA class, as all mRNAs have a few similar properties and functions. In OOP, any individual mRNA with an individual gene and sequence is an instance or object of class mRNA. In Python, integer, string, list, dictionary, and everything else constitute the "class". That is precisely why when we use the type() function, we get the word "class" - implicating the class it belongs to. Functions within the class are called methods, and we have seen various methods in the previous sections.

We first need to define a class using the keyword "class" to build our own custom object in Python. Consider creating objects to reflect the details of mRNA. Every object will be a single mRNA object. First, we have to create a class. Let us create a class named mRNA, which is an empty class, as shown below:

Code:
```
class Mrna:
    pass
```

Here, the "pass" keyword is used to show that the class is empty.
__int__() method
__int__() is a special method that initializes an object. This method is automatically executed every time a class object is created. The __int__() method is typically used for operations required before the object is produced.

Code:
```
class Mrna:


    # initializer
    def __init__(self):
        pass
```

When __int__() is specified in a class description, it should have the "self" parameter as its first parameter or variable. The "self" parameter refers to the object itself. This is to retrieve or set variables of the specific instance. This variable or argument does not always have to be named "self". We can call it whatever we want, but this is the common procedure.

Two key elements of a class are attributes and methods. Attributes are characteristics of objects. For mRNA, these will be like accession number, sequence, etc. These attributes are named as variables while defining a class, and every object can have its own values in variables. Attributes can be of two types: instance attributes and class attributes.

Code:
```
class Mrna:
# class attribute
contains = 'Nucleic Acid'

# initializer with instance attributes
def __init__(self, gene_name, seq):
self.gene_name = gene_name
self.seq = seq
```

Class attributes are common for all objects, whereas instance attributes are unique objects of a class. In the next example, we will define a class and then create an object:

Code:
```
# Defining a class
class Mrna:

    # class attribute
```

```
            contains = 'Nucleic Acid'

        # initializer with instance attributes
      def __init__(self, gene_name, seq):
           self.gene_name = gene_name
           self.seq = seq

    # Creating an object
    mRNA_1 = Mrna('IGKJ1','GUGGACGUUCGGCCAAGGGACCAAGGUGGAAAUCAAAC')
```

Output:
```
<__main__.Mrna at 0x20ba300ce80>
```

The output shows that the object has been created from class at a certain location.

Instance Methods

Instance methods are the functions defined inside a class for its objects. These functions can take the attributes of a class as arguments. Such functions can be called on objects as a methods or function of the object using the "." dot operator. Let us review the next example:

Code:
```
class Mrna:
   # class attribute
   contains = 'Nucleic Acid'

   # initializer with instance attributes
   def __init__(self, gene_name, seq):
       self.gene_name = gene_name
       self.seq = seq

   # method 1
   def Details(self):
       print("Gene Name is", self.gene_name,'and sequence is', self.seq)

   # method 2
   def cDNA(self):
     print('Its cDNA is', self.seq.replace('U','T'))

mRNA_1 = Mrna('IGKJ1','GUGGACGUUCGGCCAAGGGACCAAGGUGGAAAUCAAAC')
mRNA_1.Details()
mRNA_1.cDNA()
```

Output:
```
Gene Name is IGKJ1 and sequence is GUGGACGUUCGGCCAAGGGACCAAGGUGGAAAUCAAAC
Its cDNA is GTGGACGTTCGGCCAAGGGACCAAGGTGGAAATCAAAC
```

Here, there are two methods: "Details", which prints the details of the object and "cDNA", which determines the cDNA for the mRNA sequence.

Now, we can relate all of the built-in methods for datatype classes, how these were built, and how these work.

File Handling in Python

While writing and executing statements and programs, we were utilizing or accessing the primary memory of our computer. As discussed in Chapter One, our computer consists of primary and secondary memory - where primary memory disappears when we turn off our system. Secondary memory, which is the hard drive, can store the memory even after the system is turned off. The files are generally stored in the hard drive and have specific paths for their locations. Biological data is also stored in files with specific formats like PDB, networks, sequence files, and others. Therefore, accessing and manipulating files are essential for a biologist. Massive data is also stored and shared through files, so it is important to learn how to read a file in Python.

The first step in file handing is opening a file with the built-in "open()" function in Python. By calling the open() function, we are asking the operating system to find the file and ensure that it exists.

Code:
```
f = open('myfile.txt') # If file is in current directory
f = open('C:\Python33\Scripts\myfile.txt') # opening file with Exact Path
```

When we simply mention the filename, it is presumed that the file is in the same folder as Python. We can also specify the exact path where the file is located, as shown in the second line of code.

Character	Mode	Description
'r'	Read (default)	Open a file for read-only
'w'	Write	Open a file for write-only (overwrite)
'a'	Append	Open a file for write-only (append)
'r+'	Read+Write	Open a file for both reading and writing
'x'	Create	Create a new file
't'	Text (default)	Read and write strings from and to the file
'b'	Binary	Read and write bytes objects from and to the file. This mode is used for all files that do not contain text (e.g. images).

Specify File Mode

There are eight types of modes for operation on files. These are the following:
Here are some examples:

Code:
```
#opening a file in default mode
f = open('myfile.txt')

#Same as opening a file in default
f = open('myfile.txt', 'rt')

#opening a file for reading and writing
f = open('myfile.txt', 'r+')
```

```
#opening a bit file to read
f = open('myfile.txt', 'rb')
```

In the example below, we will open a file named "myfile.txt", which is present in the current directory or folder, and read it. The content of the file is:

> First line of the file.
> Second line of the file.
> Third line of the file.

Code:
```
f = open('my_file.txt')
print(f.read())
```

Output:
```
First line of the file.
Second line of the file.
Third line of the file.
```

In this example, "open()" is a class which deals with the operations of files. The "open" mode is initiated with the name or location of the file. This class has a method called read(), which reads all of the contents of the file. Open handler has a method for returning the list of all of the lines known as the readlines().

Code:
```
f = open('my_file.txt')
print(f.readlines())
```

Output:
```
['First line of the file.\n', 'Second line of the file.\n', 'Third line of
the file.']
```

As indicated in the output, the list of the string contains a backspace sequence "\n" in first two of its items which is for new line characters. We can get rid of it by using the split() method of strings.

We can write inside a file by opening the file in write mode and using the write() method of the handler class open().

Code:
```
f = open('my_new_file.txt', 'w')
f.write('My first line of the file.')

lines = ['New line 1\n', 'New line 2\n', 'New line 3']
f.writelines(lines)
```

The new file will be saved and can be found in the current folder or directory. The contents of the file will be as follow:

My first line of the file. New line 1
New line 2
New line 3

Once we open a file, we need to close it with the close() methods of the "f" object. The f.close() will close the file.

Handling CSV files

A CSV file is a delimited text file that uses commas to distinguish values. CSV file is used for storing tabular data. The Python CSV built-in library allows reading, writing, and processing data from and into CSV files. The name of the CSV file we are using is "diabetes.csv" and the file contains:

Pregnancies,Glucose,BloodPressure,SkinThickness,Insulin,BMI,DiabetesPedigreeFunction,
Age,Outcome

6,148,72,35,0,33.6,0.627,50,1
1,85,66,29,0,26.6,0.351,31,0
8,183,64,0,0,23.3,0.672,32,1
1,89,66,23,94,28.1,0.167,21,0
0,137,40,35,168,43.1,2.288,33,1
5,116,74,0,0,25.6,0.201,30,0
3,78,50,32,88,31,0.248,26,1
10,115,0,0,0,35.3,0.134,29,0
2,197,70,45,543,30.5,0.158,53,1
8,125,96,0,0,0,0.232,54,1

Let us use Python's CSV library for reading the values in a CSV file.

Code:

```
import csv

with open('diabetes.csv') as f:
    reader = csv.reader(f)
    for row in reader:
        print(row)
```

Output:

```
['Pregnancies', 'Glucose', 'BloodPressure', 'SkinThickness', 'Insulin',
'BMI', 'DiabetesPedigreeFunction', 'Age', 'Outcome']
['6', '148', '72', '35', '0', '33.6', '0.627', '50', '1']
['1', '85', '66', '29', '0', '26.6', '0.351', '31', '0']
['8', '183', '64', '0', '0', '23.3', '0.672', '32', '1']
['1', '89', '66', '23', '94', '28.1', '0.167', '21', '0']
['0', '137', '40', '35', '168', '43.1', '2.288', '33', '1']
['5', '116', '74', '0', '0', '25.6', '0.201', '30', '0']
['3', '78', '50', '32', '88', '31', '0.248', '26', '1']
['10', '115', '0', '0', '0', '35.3', '0.134', '29', '0']
['2', '197', '70', '45', '543', '30.5', '0.158', '53', '1']
['8', '125', '96', '0', '0', '0', '0.232', '54', '1']
```

Here we have used the "with" statement so that we do not have to write f.close() every time we open the file. Once the program comes out of the "with" block, the file closes automatically. Here the csv.read () method returns a list of values in a nested list format, as discussed in nested list section of this chapter.

We can write a list of items into CSV format using the csv.write() method.

Code:

```
with open('my_new_csvfile.csv', 'w') as f:
writer = csv.writer(f)
writer.writerow(['8', '183', '64', '0', '0', '23.3', '0.672', '32', '1'])
writer.writerow(['6', '148', '72', '35', '0', '33.6', '0.627', '50', '1'])
```

A new file named "my_new_csvfile.csv" will be created in the current directory.

The CSV library can also be used for reading and writing tab-separated files. Additionally, we can create a dictionary using the values of a CSV file.

Therefore, the basics of Python required for data analysis ends here. In the next chapter, we shall dive deeper into biological data and learn to use Biopython, which will help in more favorable and convenient working conditions.

Exercise

1. What are Booleans, and how these are associated with conditional statements?

2. Mention some of the best practices for naming a variable in Python.

3. Given a DNA sequence: "TGGGACAAGGGGTCACCCGAGTGCTGTCTTCCAATCTACTT"
   If this DNA sequence is digested by a restriction enzyme whose recognition site in "CCC", find the resulting fragments using Python programming.

4. Given this nested dictionary, use indexing to grab all "G2".
   {'nodes':['G1', 'G2', 'G3'], edges:[{'id':1, 'source': 'G1', 'end': 'G3'},{'id':2, 'source': 'G3', 'end': 'G2'}]}

5. What type of datatypes cannot be used as a key in the Python dictionary?

6. Write a function to retrieve common items from two lists.

7. Create a class for DNA with a sequence and add a method that will produce its complementary string.

3

Biopython

Introduction

Biophython is a set of libraries for biological data analysis written in Python by an international team of developers. Learning the application Python in biological sciences will be incomplete if we do not introduce the Biopython package to readers. It was in 1999 when the Biopython Project was formed as a collaboration to develop open-source bioinformatics tools, and it was first released in July 2000. Since then, it has evolved into a huge range of bioinformatics modules and scripts which can easily be accessed on biopython.org. The purpose of Biopython is to make the use of Python for bioinformatics as simple as possible by providing modules and scripts frequently required for bioinformatics applications. It provides parsers with several file-formats including FASTA, GenBank, SwissProt, PDB, etc. It also offers access to online resources and biological databases with interfaces for many standard tools like BLAST, Clustal W alignment program, sophisticated clustering methods, and several other services and tools for making the tasks of bioinformatics simpler and quicker. In this chapter, we will witness the simplicity of using Biopython for solving bioinformatics problems. The official guide book for Biopython is available on http://biopython.org/DIST/docs/tutorial/Tutorial.pdf.

Installing Biopython

Installing Biopython is the same as installing any package in Python. We can install it by using either the "conda" or "pip" package installer in the Anaconda command prompt. Once the package is installed, we can test it by opening a Jupyter Notebook and running "Import Bio' to import libraries and methods from the Biopython package.

Code:
```
import Bio
print(Bio.__version__)
```
Output:
```
1.72
```

If the code above is executed properly without any errors, then this implies the successful installation of Biopython. The "print(Bio.__version__)" statement is used to check the version of Biopython. Note that the "__" before and after "version" are two underscores "_" consecutively.

Biopython Seq Class

In the classes and objects section of the previous chapter, we have created a class called "Mrna" which has two instance methods, "Details" and "cDNA" - which could do simple operations on our mRNA object. Likewise, Biopython also has a class called "seq", which is like strings in Python but contains various other complex operations on biological sequences. Let us study a few examples below:

Code:

```
from Bio.Seq import Seq
my_seq = Seq("ATGGCCTTAAA")
print(type(my_seq))
```

Output:

```
<class 'Bio.Seq.Seq'>
```

In using the Biopython Seq class, we have to import the class from the Biopython package using the "from" and "import" keywords - as shown in the first line of the code example above. Next, we can create a seq object by using a string as an argument, as done in the second statement of the afore-mentioned example. We can see that our "Bio. Seq" object is created. With the creation of the seq object, Biopython has built various instance methods that are listed below, and these can be used on the sequence object like "complement", "transcribe", "translate", etc. Biologists can easily recognize the terms and infer what these methods do. An important thing to note here is that we have to instruct Biopython about the biological type of the sequence, like DNA, RNA, or protein. Defining the type of sequence can help Biopython to figure out which methods are applicable to the sequence. The type of sequence is stored and known as the "alphabet" in the Biopython seq object. For instance, the English language has 26 alphabets; DNA and RNA have 4 alphabets; whereas protein has 20 alphabets.

Code:

```
from Bio.Alphabet import generic_dna, generic_protein, generic_rna
my_seq = Seq("ATGGCCTTAAA")
dna = Seq("ATGGCCTTAAA", generic_dna)
rna = Seq("ACCCGUUGU", generic_rna)
protein = Seq("AKKKGGGUUULL", generic_protein)
print(my_seq.alphabet)
print(dna.alphabet)
print(rna.alphabet)
print(protein.alphabet)
```

Output:
```
Alphabet()
DNAAlphabet()
RNAAlphabet()
ProteinAlphabet()
```

The alphabet object is essential, making the seq object much more than a string. For example, the sequence ATGGCCTTAAA is a DNA sequence when we define the DNA alphabet with it, rather than a protein sequence. When we initiate a seq object without an alphabet argument, then it is initiated with a global alphabet object which does not help Biopython in knowing the properties of the sequence. It is always recommended to initiate a seq object with an alphabet argument. Let us study some of the instance methods of seq objects in the following example:

Code:

```
# Central dogma example.
gene = Seq("ATGGCCTTAAAT", generic_dna)
#transcription
transcript = gene.transcribe()
print(transcript)
```

```
print(transcript.alphabet)
#Translation
protein = transcript.translate()
print(protein)
print(protein.alphabet)
```

Output:

```
AUGGCCUUAAAU
RNAAlphabet()
MALN
ExtendedIUPACProtein()
```

The central dogma broadly comprises the principle of transcribing DNA to RNA and, consequently, translating RNA into protein. Biopython itself can change the alphabet object for mRNA and protein once we provide the argument as generic_dna while initiating a DNA seq object. We can directly transcribe a DNA sequence to an RNA sequence using the ".transcribe()" method of the seq object and translate an RNA sequence into protein using the "translate" method. Sequences or strings in seq objects are indexed, just like strings objects, and can be sliced using square bracket notation. Seq objects also support "+" and "*" operations like string objects.

Parsing Sequence Files

Bio. Seq objects are used for the representation of biological sequences. However, we often require storing some information along with the sequence, such as the ID, description, and taxonomy. To do this, the Biopython package has a "SeqRecord" class above the "Seq" Class, which has more sophisticated features like storing various identifiers and annotations associated with the sequence or the seq class.

Code:

```
from Bio.Seq import Seq
from Bio.SeqRecord import SeqRecord
from Bio.Alphabet import generic_protein
protein_record = SeqRecord(Seq("MRAKWRKKRMRRLKRKRRKMRQRSK",
generic_protein),
id="P62945", name="RL41_HUMAN",
description="60S ribosomal protein L41")
print(protein_record)
```

Output:

```
ID: P62945
Name: RL41_HUMAN
Description: 60S ribosomal protein L41
Number of features: 0
Seq('MRAKWRKKRMRRLKRKRRKMRQRSK', ProteinAlphabet())
```

We can create a "SeqRecord" object by giving various arguments, as shown above, although it is designed to acquire all the information about the biological sequence from stored or fetched files in FASTA, Genbank, SwissProt, and various other file formats. We already know how to read and write files stored in our hard drive using a Python handler. In BioPython, we have a "SeqIO" module that can read and write sequence files of various formats and make our programming experience smoother. For example, the FASTA format is a text-based representation of nucleotide or protein sequences, where the first line begins with a ">" symbol and the consecutive lines contain the sequence in a single-letter representation of nucleotides or amino acids. Each line can contain 80 to 120 characters, but generally, it is preferred to keep a character limit at less than 80 per line. Here is an example of a FASTA-formatted file.

```
>sp|Q9SE35|20-107
QSIADLAAANLSTEDSKSAQLISADSSDDASDSSVESVDAASSDVSGSSVESVDVSGSSL
ESVDVSGSSLESVDDSSEDSEEEELRIL
```

Let us bring this above protein FASTA file into a SeqRecord object without using a Biopython parser.

Code:

```
from Bio.Seq import Seq
from Bio.SeqRecord import SeqRecord
from Bio.Alphabet import generic_protein
sequence = ''
with open('sample.fasta','r') as f:
lines = f.readlines() # read the lines of the file
for line in lines:
    if line.startswith('>'):
        description = line.rstrip() # to get rid of the newline character
    else:
        sequence = sequence + line.rstrip()
protein_record = SeqRecord(Seq(sequence,generic_protein),
id=description , name=description ,
description=description)
print(protein_record)
```

Output:

```
ID: >sp|Q9SE35|20-107
Name: >sp|Q9SE35|20-107
Description: >sp|Q9SE35|20-107
Number of features: 0
Seq('QSIADLAAANLSTEDSKSAQLISADSSDDASDSSVESVDAASSDVSGSSVESVD...RIL',
ProteinAlphabet())
```

We read the lines of the file using Python's built-in file object and handler. Next, we iterate over each of the lines and insert a line, which starts with the ">" symbol into the description variable. Subsequent lines are added with an empty string variable named as a sequence. The "rstrip()" method is used to get rid of the newline character "\n", which we usually get at the end of lines.

This code seems to be easy, although we had to write some statements. The same problem will become a bit more complicated once we deal with several sequences in a single FASTA file. The problem will turn even more complicated when we try to parse a GenBank format file. An example of a GenBank format file is shown in Figure 3.1 below:

```
LOCUS           X03109                  1822 bp    DNA      linear    PRI 14-NOV-2006
DEFINITION   Chimpanzee fetal G-gamma-globin gene.
ACCESSION    X03109
VERSION      X03109.1
KEYWORDS     direct repeat; G-gamma-globin; gamma-globin; tandem repeat.
SOURCE       Pan troglodytes (chimpanzee)
  ORGANISM   Pan troglodytes
             Eukaryota; Metazoa; Chordata; Craniata; Vertebrata; Euteleostomi;
             Mammalia; Eutheria; Euarchontoglires; Primates; Haplorrhini;
             Catarrhini; Hominidae; Pan.
REFERENCE    1  (bases 1 to 1822)
  AUTHORS    Slightom,J.L., Chang,L.Y., Koop,B.F. and Goodman,M.
  TITLE      Chimpanzee fetal G gamma and A gamma globin gene nucleotide
             sequences provide further evidence of gene conversions in hominine
             evolution
  JOURNAL    Mol. Biol. Evol. 2 (5), 370-389 (1985)
  PUBMED     3870867
REFERENCE    2  (bases 1 to 1822)
  AUTHORS    Slightom,J.L.
  TITLE      Direct Submission
  JOURNAL    Submitted (07-JUL-1986)
FEATURES             Location/Qualifiers
     source          1..1822
                     /organism="Pan troglodytes"
                     /mol_type="genomic DNA"
                     /db_xref="taxon:9598"
     regulatory      24..28
                     /regulatory_class="TATA_box"
     mRNA            join(55..199,322..544,1438..1652)
                     /product="G-gamma-globin"
ORIGIN
        1 ccggcggctg gctagggatg aagaataaaa ggaagcaccc tccagcagtt ccacacactc
       61 gcttctggaa cgtctgaggt tatcaataag ctcctagtcc agacgccatg ggtcatttca
      121 cagaggagga caaggctact atcacaagcc tgtgggcaa ggtgaatgtg gaagatgctg
      181 gaggagaaac cctgggaagg taggctctgg tgaccaggac aagggaggga aggaaggacc
      241 ctgtgcctgg caaaagtcca ggtcacttct caggatctgt ggcaccttct gactgtcaaa
      301 ctgttcttgt caatcttaca ggctcctggt tgtctaccca tggacccaga ggttctttga
      361 cagctttggc aacctgtcct ctgcctctgc catcatgggc aaccccaagg tcaaggcaca
      421 tggcaagaag gtgctgactt ccttgggaga tgccataaag cacctggatg atctcaaggg
                     /number=2
```

FIGURE 3.1 GenBank Format File Example.

We can recognize that there is a significant amount of descriptions and features in the GenBank formatted file. Here comes Biopython to rescue us from these complications. The "SeqIO" module of Biopython helps us a lot in dealing with these complex file formats. Let us study how to use the "SeqIO" module to parse our FASTA file:

Code:

```
from Bio import SeqIO
records = SeqIO.parse("sample.fasta", "fasta")
for record in records:
    print(record)
```

Output:

```
ID: sp|Q9SE35|20-107
Name: sp|Q9SE35|20-107
Description: sp|Q9SE35|20-107
Number of features: 0 Seq('QSIADLAAANLSTEDSKSAQLISADSSDDASDSSVESVDAASSDV-
SGSSVESVD…RIL', SingleLetterAlphabet())
```

We can determine that the output is the same as the previous output, but it has fewer statements. Although it is good to write a program from scratch to be able to learn and understand the code and concepts, we can also use packages and methods to make our program more stable and efficient. Python is popular primarily because of the availability of ready-to-use modules for implementing various modern technologies. The "parse" method in the "SeqIO" model takes the file name or path and format of the file as arguments and returns an iterator of SeqRecods. Items in this iterator can be accessed by a "for" loop. In the example above, we have only one sequence, but we can use the same method for a FASTA file with multiple sequences. One such example with the file name "multi_sequence.fasta" is illustrated below.

```
>sp|Q3ZM63|ETDA_HUMAN
MDKEVPKGSPREPALNIKKSDKSFKRKKPTENVLIF
LINRQLGRHRSDIDLSRWVWMLS
>sp|P53803|RPAB4_HUMAN
MDTQKDVQPPKQQPMIYICGECHTENEIKSRDPIRC
RECGYRIMYKKRTKRLVVFDAR
>sp|Q538Z0|LUZP6_HUMAN
MKSVISYALYQVQTGSLPVYSSVLTKSPLQLQTVIY
RLIVQIQHLNIPSSSSTHSSPF
>sp|Q9BZ97|TTY13_HUMAN
MKTQDDGVLPPYDVNQLLGWDLNLSLFLGLCLMLLL
AGSCLPSPGITGLSHGSNREDR
>sp|P58511|SI11A_HUMAN Small
MNWKVLEHVPLLLYILAAKTLILCLTFAGVKMYQRK
RLEAKQQKLEAERKKQSEKKDN
```

Code:

```python
from Bio import SeqIO
record_list = []
for record in SeqIO.parse("multi_sequence.fasta", "fasta"):
    print(record.id)
    record_list.append(record)
```

Output:

```
sp|Q3ZM63|ETDA_HUMAN
sp|P53803|RPAB4_HUMAN
sp|Q538Z0|LUZP6_HUMAN
sp|Q9BZ97|TTY13_HUMAN
sp|P58511|SI11A_HUMAN
```

We can create a list of SeqRecords by appending the items from the SeqIO.parse's iterator to an empty list. We can also acquire a list of SeqRecods by inputting this iterator as an argument to the "list ()" function. SeqIO can deal with various biological sequence file formats like FASTA, GenBank, EMBL, and SwissProt, and it can also read the output files of various NGS platforms like Solexa and

Illumina FASTQ files. Let us read a GenBank format file which is loaded with many descriptions, annotations, and/or features. The GenBank file used in the code below is provided with the practice material of this chapter as "sequence.gb", or it can be downloaded from NCBI: https://www.ncbi.nlm. nih.gov/sviewer/viewer.cgi?tool=portal&save=file&log$=seqview&db=nuccore&report=gbwithparts& id=38219&withparts=on&showgi=1

Code:

```
from Bio import SeqIO
records = SeqIO.parse("sequence.gb","genbank")
for record in records:
    print(record)
```

Output:

```
ID: X03109.1
Name: X03109
Description: Chimpanzee fetal G-gamma-globin gene
Number of features: 13
/molecule_type=DNA
/topology=linear
/data_file_division=PRI
/date=14-NOV-2006
/accessions=['X03109']
/sequence_version=1
/gi=38219
/keywords=['direct repeat', 'G-gamma-globin', 'gamma-globin', 'tandem
repeat']
/source=Pan troglodytes (chimpanzee)
/organism=Pan troglodytes
/taxonomy=['Eukaryota', 'Metazoa', 'Chordata', 'Craniata', 'Vertebrata',
'Euteleostomi', 'Mammalia', 'Eutheria', 'Euarchontoglires', 'Primates',
'Haplorrhini', 'Catarrhini', 'Hominidae', 'Pan']
/references=[Reference(title='Chimpanzee fetal G gamma and A gamma globin
gene nucleotide sequences provide further evidence of gene conversions in ho-
minine evolution', …), Reference(title='Direct Submission', …)] Seq('CCGGCGG
CTGGCTAGGGATGAAGAATAAAAGGAAGCACCCTCCAGCAGTTCCAC...AAT',
IUPACAmbiguousDNA())
```

As previously mentioned, the GenBank file format contains a huge amount of features -along with the ID, name, description, and "seq" object. The "SeqIO.parse" method takes the small case file format as an argument along with the file name. The details about various formats and their arguments' names can be found in Biopython's documentation, and readers are advised to go through it.

Writing Files

In the process of writing sequence files from SeqRecords in various formats, Biopython has "SeqIO.write()" method. This method generally takes three arguments: the first is the SeqRecord or a list of SeqRecords; next is the output file name; last is the format we choose for the output file.

Code:

```
from Bio import SeqIO
from Bio.Seq import Seq
from Bio.SeqRecord import SeqRecord
from Bio.Alphabet import generic_protein
protein = SeqRecord(Seq("MRAKWRKKRMRRLKRKRRKMRQRSK",
generic_protein),
id="P62945", name="RL41_HUMAN",
description="60S ribosomal protein L41")
SeqIO.write(protein, "example.fasta", "fasta")
with open('example.fasta') as f:
    print(f.read())
```

Output:

```
>P62945 60S ribosomal protein L41
MRAKWRKKRMRRLKRKRRKMRQRSK
```

The output is a FASTA file named "example.fasta". In this way, we can write a single sequence containing the FASTA file from a single SeqRecord object. To be able to write a FASTA file containing multiple sequences, we require a list of SeqRecords.

Code:

```
from Bio import SeqIO
from Bio.Seq import Seq
from Bio.SeqRecord import SeqRecord
from Bio.Alphabet import generic_protein
record_list = []
protein_record_1 = SeqRecord(Seq("MRAKWRKKRMRRLKRKRRKMRQRSK",
generic_protein),
id="P62945", name="RL41_HUMAN",
description="60S ribosomal protein L41")
protein_record_2 = SeqRecord(Seq("MGINTRELFLNFTIVLITVILMWLLVRSYQY",
generic_protein),
id="O00631", name="SARCO_HUMAN",
description="Sarcolipin")
record_list = [protein_record_1,protein_record_2]
SeqIO.write(record_list, "example_2.fasta", "fasta")
with open('example_2.fasta') as f:
    print(f.read())
```

Output:

```
>P62945 60S ribosomal protein L41
MRAKWRKKRMRRLKRKRRKMRQRSK
>O00631 Sarcolipin
MGINTRELFLNFTIVLITVILMWLLVRSYQY
```

In the same manner, we can write the SeqRecords in various formats, and these can be found in Biopython documentation.

Pairwise Sequence Alignment

The Pairwise Sequence Alignment method involves aligning two sequences of DNA, RNA, or proteins to find the regions of similarity among them. Pairwise Sequence Alignments are used for identifying similar characteristics, structures, or evolutionary relationships. In most cases, researchers align two protein sequences to determine associations, such as homology. There are various ways to align two sequences and deduce their characteristics, such as:

1. The dot-matrix method is a process where two sequences can be aligned using a two-dimensional matrix, and points are plotted at points of identity between the x-axis and the y-axis. Therefore, we can visualize the plots and infer the results. Since there could be various random matches in the case of entire genome sequences, the dot-matrix method of alignment is only suitable for highly similar and short sequences.

2. Another method of Pairwise Sequence Alignment is the dynamic programming approach. The basis of the dynamic programming approach for sequence alignment is the scoring method - where the score indicates the measure of similarity between two sequences. This algorithm of alignment generates a matrix of numbers that represents all possible alignments between the sequences. The score is then used as a measure of comparison between two sequences. Three basic factors are taken into account when determining if the scores are a match, mismatch, or if there is a gap penalty. The dynamic programming approach of alignment is the most accurate method for sequence alignment. However, its drawback is that it is extremely slow and is computationally intensive for large sequences. There are two types of alignments:

   a. The global alignment method takes the whole length of the two sequences into consideration and tries to match them to get the best alignment in the entire sequence. It is obtained by inserting gaps or spaces in both of the sequences. The Needleman-Wunsch algorithm is one of the algorithms that use dynamic programming to produce a global alignment. It was the first and best application of dynamic programming in biological sequence alignment.

   b. The local alignment method, in contrast to global alignment, provides the most similar subsequences between the two sequences. The Smith-Waterman algorithm is one of the implementations of the local alignment algorithm that applies dynamic programming. This algorithm is similar to Needleman-Wunsch, but the scoring mechanism varies slightly.

In this section, we will observe the application of dynamic Pairwise Sequence Alignment algorithms using Biopython.

There is a built-in module in Biopython called "Bio. Align" for sequence alignment. This "Bio.Align" has a "PairwiseAligner" class for executing Pairwise Sequence Alignment using dynamic programming. Let us accomplish a simple nucleotide alignment in Biopython below:

Code:

```
from Bio import Align
aligner = Align.PairwiseAligner()
seq1 = 'TACGCCCGC'
seq2 = 'TAGCCCATGC'
results = aligner.align(seq1, seq2)
for result in results:
    print(result)
score = aligner.score(seq1, seq2)
print(score)
```

Output:

```
TACGCCC--GC
||-||||--||
TA-GCCCATGC
8.0
```

By default, the PairwiseAligner does global alignment. In the snap of code illustrated above, we are creating an object of "Align.PairwiseAligner()" as the "aligner". This object has various default parameters - like the match, mismatch, and gap penalties - and we can also change the mode from global to local alignment. Here is a way to view the paraments and their default values of this object:

Code:

```
from Bio import Align
aligner = Align.PairwiseAligner()
print(aligner)
print('----------------')
print(aligner.algorithm)
```

Output:

```
Pairwise sequence aligner with parameters
match_score: 1.000000
mismatch_score: 0.000000
  .
  .
  .
  .
query_right_extend_gap_score: 0.000000
mode: global
----------------
Needleman-Wunsch
```

We can customize the scores for the match, mismatch, and various gap penalties. An example is shown below:

Code:

```
from Bio import Align
aligner = Align.PairwiseAligner()
seq1 = 'TACGCCCGC'
seq2 = 'TAGCCCATGC'
aligner.match_score = 1.0
aligner.mismatch_score = -2.0
aligner.gap_score = -2.5
alignments = aligner.align(seq1, seq2)
for alignment in alignments:
    print(alignment)
score = aligner.score(seq1, seq2)
print(score)
```

Output:

```
TACGCCC--GC
||-||||--||
TA-GCCCATGC
0.5
```

In the example above, we have changed the match, mismatch, and gap score - therefore, leading to a change in the overall score, which was formerly 8.0. Now, it is 0.5. Increasing the "mismatch score" and "gap score" will make the alignment more stringent. When doing local alignment, we need to change the "mode" parameter to "local".

Code:

```
from Bio import Align
aligner = Align.PairwiseAligner()
aligner.mode = 'local'
seq1 = 'TACGCCCGC'
seq2 = 'TAGCCCATGC'
alignments = aligner.align(seq1, seq2)
for alignment in alignments:
    print(alignment)
score = aligner.score(seq1, seq2)
print(score)
```

Output:

```
TACGCCC--GC
||-||||--||
TA-GCCCATGC
8.0
```

Let us put some stringency to the various scoring parameters to study the difference in the alignment and score.

Code:

```
from Bio import Align
aligner = Align.PairwiseAligner()
aligner.mode = 'local'
seq1 = 'TACGCCCGC'
seq2 = 'TAGCCCATGC'
aligner.match_score = 1.0
aligner.mismatch_score = -2.0
aligner.gap_score = -2.5
alignments = aligner.align(seq1, seq2)
for alignment in alignments:
    print(alignment)
score = aligner.score(seq1, seq2)
print(score)
```

Output:

```
TACGCCCGC
| | | |
TAGCCCATGC
4.0
```

After putting some stringency in the scores, we can note the local alignment in the subsequences. The examples given above are the alignments of nucleotide sequences. We will also regard the alignment of the protein sequence; however, before doing so, we will discuss substitution matrixes. Substitution matrices are matrices of similarity between two amino acids or nucleotides. In the case of protein sequence alignment, if we find the same amino acid in a particular location/position, then we will put the identity score as straightforward. On the other hand, in the case of mismatch, we need to determine similarities in amino acids for proper biological interpretation. This is because the amino acids have physical, chemical, and geometric properties, and the mismatches can be classified into "similarity" and "non-similarity". For example, if a hydrophobic amino acid replaces another hydrophobic amino acid, then the change may be tolerable, and such substitutions are known as conservative substitutions. Under other conditions, if the hydrophobic amino acid is replaced by hydrophilic amino acid, and this change can not be sustained, then such substitution is considered to be radical. Other scoring matrices can be developed by calculating the frequency of amino acid substitutions for homologous proteins in different spices. Evolution has proven to give preference to some types of amino acid substitutions over others. Therefore, these substitutions can be given higher scores. Biopython has about 25 substitution matrices, and these can be retrieved from "Bio. Align import substitution_matrices".

One of the most common substitution matrices is "BLOSUM62". Let us load the substitution matrix "BLOSUM62" using Biopython (Figure 3.2).

Code:

```
from Bio.Align import substitution_matrices
matrix = substitution_matrices.load("BLOSUM62")
print(matrix)
```

```
#  Matrix made by matblas from blosum62.iij
#  * column uses minimum score
#  BLOSUM Clustered Scoring Matrix in 1/2 Bit Units
#  Blocks Database = /data/blocks_5.0/blocks.dat
#  Cluster Percentage: >= 62
#  Entropy =   0.6979, Expected =  -0.5209
      A    R    N    D    C    Q    E    G    H    I    L    K    M    F    P    S    T    W    Y    V    B    Z    X    *
A   4.0 -1.0 -2.0 -2.0  0.0 -1.0 -1.0  0.0 -2.0 -1.0 -1.0 -1.0 -1.0 -2.0 -1.0  1.0  0.0 -3.0 -2.0  0.0 -2.0 -1.0  0.0 -4.0
R  -1.0  5.0  0.0 -2.0 -3.0  1.0  0.0 -2.0  0.0 -3.0 -2.0  2.0 -1.0 -3.0 -2.0 -1.0 -1.0 -3.0 -2.0 -3.0 -1.0  0.0 -1.0 -4.0
N  -2.0  0.0  6.0  1.0 -3.0  0.0  0.0  0.0  1.0 -3.0 -3.0  0.0 -2.0 -3.0 -2.0  1.0  0.0 -4.0 -2.0 -3.0  3.0  0.0 -1.0 -4.0
D  -2.0 -2.0  1.0  6.0 -3.0  0.0  2.0 -1.0 -1.0 -3.0 -4.0 -1.0 -3.0 -3.0 -1.0  0.0 -1.0 -4.0 -3.0 -3.0  4.0  1.0 -1.0 -4.0
C   0.0 -3.0 -3.0 -3.0  9.0 -3.0 -4.0 -3.0 -3.0 -1.0 -1.0 -3.0 -1.0 -2.0 -3.0 -1.0 -1.0 -2.0 -2.0 -1.0 -3.0 -3.0 -2.0 -4.0
Q  -1.0  1.0  0.0  0.0 -3.0  5.0  2.0 -2.0  0.0 -3.0 -2.0  1.0  0.0 -3.0 -1.0  0.0 -1.0 -2.0 -1.0 -2.0  0.0  3.0 -1.0 -4.0
E  -1.0  0.0  0.0  2.0 -4.0  2.0  5.0 -2.0  0.0 -3.0 -3.0  1.0 -2.0 -3.0 -1.0  0.0 -1.0 -3.0 -2.0 -2.0  1.0  4.0 -1.0 -4.0
G   0.0 -2.0  0.0 -1.0 -3.0 -2.0 -2.0  6.0 -2.0 -4.0 -4.0 -2.0 -3.0 -3.0 -2.0  0.0 -2.0 -2.0 -3.0 -3.0 -1.0 -2.0 -1.0 -4.0
H  -2.0  0.0  1.0 -1.0 -3.0  0.0  0.0 -2.0  8.0 -3.0 -3.0 -1.0 -2.0 -1.0 -2.0 -1.0 -2.0 -2.0  2.0 -3.0  0.0  0.0 -1.0 -4.0
I  -1.0 -3.0 -3.0 -3.0 -1.0 -3.0 -3.0 -4.0 -3.0  4.0  2.0 -3.0  1.0  0.0 -3.0 -2.0 -1.0 -3.0 -1.0  3.0 -3.0 -3.0 -1.0 -4.0
L  -1.0 -2.0 -3.0 -4.0 -1.0 -2.0 -3.0 -4.0 -3.0  2.0  4.0 -2.0  2.0  0.0 -3.0 -2.0 -1.0 -2.0 -1.0  1.0 -4.0 -3.0 -1.0 -4.0
K  -1.0  2.0  0.0 -1.0 -3.0  1.0  1.0 -2.0 -1.0 -3.0 -2.0  5.0 -1.0 -3.0 -1.0  0.0 -1.0 -3.0 -2.0 -2.0  0.0  1.0 -1.0 -4.0
M  -1.0 -1.0 -2.0 -3.0 -1.0  0.0 -2.0 -3.0 -2.0  1.0  2.0 -1.0  5.0  0.0 -2.0 -1.0 -1.0 -1.0 -1.0  1.0 -3.0 -1.0 -1.0 -4.0
F  -2.0 -3.0 -3.0 -3.0 -2.0 -3.0 -3.0 -3.0 -1.0  0.0  0.0 -3.0  0.0  6.0 -4.0 -2.0 -2.0  1.0  3.0 -1.0 -3.0 -3.0 -1.0 -4.0
P  -1.0 -2.0 -2.0 -1.0 -3.0 -1.0 -1.0 -2.0 -2.0 -3.0 -3.0 -1.0 -2.0 -4.0  7.0 -1.0 -1.0 -4.0 -3.0 -2.0 -2.0 -1.0 -2.0 -4.0
S   1.0 -1.0  1.0  0.0 -1.0  0.0  0.0  0.0 -1.0 -2.0 -2.0  0.0 -1.0 -2.0 -1.0  4.0  1.0 -3.0 -2.0 -2.0  0.0  0.0  0.0 -4.0
T   0.0 -1.0  0.0 -1.0 -1.0 -1.0 -1.0 -2.0 -2.0 -1.0 -1.0 -1.0 -1.0 -2.0 -1.0  1.0  5.0 -2.0 -2.0  0.0 -1.0 -1.0  0.0 -4.0
W  -3.0 -3.0 -4.0 -4.0 -2.0 -2.0 -3.0 -2.0 -2.0 -3.0 -2.0 -3.0 -1.0  1.0 -4.0 -3.0 -2.0 11.0  2.0 -3.0 -4.0 -3.0 -2.0 -4.0
Y  -2.0 -2.0 -2.0 -3.0 -2.0 -1.0 -2.0 -3.0  2.0 -1.0 -1.0 -2.0 -1.0  3.0 -3.0 -2.0 -2.0  2.0  7.0 -1.0 -3.0 -2.0 -1.0 -4.0
V   0.0 -3.0 -3.0 -3.0 -1.0 -2.0 -2.0 -3.0 -3.0  3.0  1.0 -2.0  1.0 -1.0 -2.0 -2.0  0.0 -3.0 -1.0  4.0 -3.0 -2.0 -1.0 -4.0
B  -2.0 -1.0  3.0  4.0 -3.0  0.0  1.0 -1.0  0.0 -3.0 -4.0  0.0 -3.0 -3.0 -2.0  0.0 -1.0 -4.0 -3.0 -3.0  4.0  1.0 -1.0 -4.0
Z  -1.0  0.0  0.0  1.0 -3.0  3.0  4.0 -2.0  0.0 -3.0 -3.0  1.0 -1.0 -3.0 -1.0  0.0 -1.0 -3.0 -2.0 -2.0  1.0  4.0 -1.0 -4.0
X   0.0 -1.0 -1.0 -1.0 -2.0 -1.0 -1.0 -1.0 -1.0 -1.0 -1.0 -1.0 -1.0 -1.0 -2.0  0.0  0.0 -2.0 -1.0 -1.0 -1.0 -1.0 -1.0 -4.0
*  -4.0 -4.0 -4.0 -4.0 -4.0 -4.0 -4.0 -4.0 -4.0 -4.0 -4.0 -4.0 -4.0 -4.0 -4.0 -4.0 -4.0 -4.0 -4.0 -4.0 -4.0 -4.0 -4.0  1.0
```

FIGURE 3.2 BLOSUM62 Amino Substitution Matrix.

Output:

Once we have loaded a matrix, we can assign it to the "Align.PairwiseAligner()" object. Below is an example of the amino-acid sequence global alignment:

Code:

```
from Bio import Align
aligner = Align.PairwiseAligner()
matrix = Align.substitution_matrices.load("BLOSUM62")
aligner.substitution_matrix = matrix
aligner.match_score = 1.0
aligner.mismatch_score = -2.0
aligner.gap_score = -2.5
protein_1 = 'MERSTQELFINFTVVLITVLLMWLLVRSYQY'
protein_2 = 'MGINTRELFLNFTIVLITVILMWLLVRSYQY'
score = aligner.score(protein_1 , protein_2)
alignments = aligner.align(protein_1 , protein_2)
print(score)
for alignment in alignments:
    print(alignment)
```

Output:

```
10.0
MERSTQELFINFTVVLITVLLMWLLVRSYQY
|...|.|||.|||.|||||.|||||||||||
MGINTRELFLNFTIVLITVILMWLLVRSYQY
```

In the example above, "|" represents matches and "." represents similar amino acids. We can also carry out local alignment by adjusting the parameters of the "aligner" object. It is a good and advisable exercise to practice by changing the parameters and matrices to determine the biological relevance of the alignments thus generated.

BLAST with Biopython

The Basic Local Alignment Search Tool (BLAST) is one of the most used sequence analysis tools in the biological research community. As the name suggests, it is a sequence-searching algorithm that can locate sequences that are similar to the query sequence from a large database of sequences. Imagine a scenario wherein a new gene has been sequenced. Looking into the sequence will not help us much to learn about the gene. The first step is to assign a putative function to the gene of interest. One way in which this could be achieved is by finding the genes with higher sequence similarity to our gene, so as to predict and derive some information about the new gene. Since dynamic programming would be time and computationally intensive for performing pairwise sequence alignment of our query sequence with each of the sequences in the database - Some "heuristic" methods - have been proposed. These are especially useful in large-scale database searches where a significant proportion of the sequences are likely to have a minimal match with the query sequence. One of the most popular "heuristic" methods is BLAST, developed by David J. Lipman and his colleagues in 1990. The algorithm of BLAST involves dividing the query sequence into small subsequences called k-mers - where k represents the number of mers or the length of the subsequence, and the value of k is usually between 3 and 10 for a query protein sequence. These small k-mers are then aligned with the sequences in the database, and hits are generated. These hits are further refined by elongating these subsequences according to the similarity between the query sequence and the corresponding sequence in the database. There are websites for querying using BLAST like NBCI BLAST (https://blast.ncbi.nlm.nih.gov/Blast.cgi). Here, we will access the web version of NCBI BLAST using the "Blast" module of Biopython. BLAST can be installed as standalone software on our computers, and a wrapper is available in the "Bio.Blast" package. However, since the BLAST package as well as the sequence database of NCBI need to be downloaded and consume a large space in the computer, then we will use the web accessibility of BLAST using the "NCBIWWW" module of Biopython. Let us practice below:

Code:

```
from Bio.Alphabet import generic_dna
from Bio.Blast import NCBIWWW
from Bio.Seq import Seq
sequence = Seq("ATTTTCTTGCTCTTGAGCTCTGGCACTTCTCTGCTGCTGTC", generic_dna)
result_handle = NCBIWWW.qblast("blastn", "nt", sequence)
```

To apply the online version of BLAST, we use the function qblast() in the Bio.Blast.NCBIWW module. qblast() takes three compulsory arguments:

1. The first argument is the type of blast, as shown in the table below:

TABLE 3.1

Types of BLAST Available in NCBIWWW.qblast().

Option	Query Type	DB Type	Comparison	Note
blastn	Nucleotide	Nucleotide	Nucleotide-Nucleotide	
blastp	Protein	Protein	Protein-Protein	
tblastn	Protein	Nucleotide	Protein-Protein	The database is translated into protein
blastx	Nucleotide	Protein	Protein-Protein	The queries are translated into protein

2. The second argument is the database in which we prefer to search our query sequence. In the code example above, we used "nt", which is the standard nucleotide database of NCBI. Other NCBI sequence databases can be found in the supplementary file
3. The third argument is the sequence type as the "Bio.Seq" object.

A typical search result and alignment of BLAST is shown in Figure 3.3.

These results can be downloaded in various formats, as shown in the figure above. "qblast()", by default, downloads the "XML" format. We can use the optional argument for obtaining other formats. The "result_handle" variable is the output format obtained after querying BLAST.

Code:

```
from Bio.Blast import NCBIXML
blast_records = NCBIXML.parse(result_handle)
for b in blast_records:
for alignment in b.alignments[:2]:
    for hsp in alignment.hsps:
        print('****Alignment****')
        print('sequence:', alignment.title)
        print('length:', alignment.length)
        print('e value:', hsp.expect)
        print(hsp.query[0:75] + '...')
        print(hsp.match[0:75] + '...')
        print(hsp.sbjct[0:75] + '...')
```

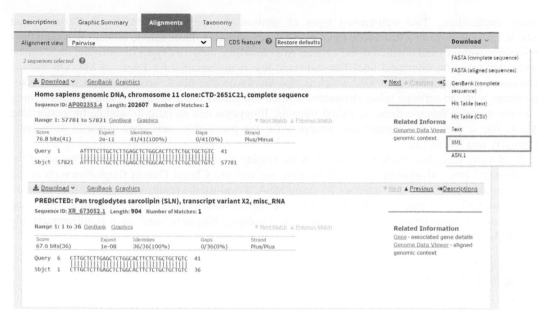

FIGURE 3.3 Search Result of Query Sequence Given in Code Example.

Output:

```
****Alignment****
sequence: gi|31790676|dbj|AP002353.4| Homo sapiens genomic DNA, chromosome
11 clone:CTD-2651C21, complete sequence
length: 202607
e value: 5.15868e-11
ATTTTCTTGCTCTTGAGCTCTGGCACTTCTCTGCTGCTGTC...
|||||||||||||||||||||||||||||||||||||||||...
ATTTTCTTGCTCTTGAGCTCTGGCACTTCTCTGCTGCTGTC...
****Alignment****
sequence: gi|694946907|ref|XR_673052.1| PREDICTED: Pan troglodytes sarco-
lipin (SLN), transcript variant X2, misc_RNA
length: 904
e value: 2.67226e-08
CTTGCTCTTGAGCTCTGGCACTTCTCTGCTGCTGTC...
|||||||||||||||||||||||||||||||||||||...
CTTGCTCTTGAGCTCTGGCACTTCTCTGCTGCTGTC...
```

As we have previously discussed, the "result_handle" variable will have the XML format results of BLAST. "Bio.Blast" has a module called NCBIXML which parses the XML-formatted results and returns the results displaying alignments.

Programmatically accessing NCBI BLAST will enable us to do a large number of searches using the "for" loops.

Multiple Sequence Alignment

Multiple sequence alignments (MSA), as the name suggests, involve alignments of more than two sequences. Multiple sequence alignment is very popular in bioinformatics sequence analysis and has many applications. Two well-known types of software for performing MSA are ClustalW and MUSCLE. These are two standalone types of software - meaning, these can be downloaded from their respective websites and installed in our system. After installing these two types of software, these can be instructed using the command line. The command line is the terminal for Linux or Mac OS and the command prompt in Windows OS, which is also known as that black window where we can communicate with the software using statements. Python can access these statements and command lines, and the modules that use them are called wappers. Biopython has no built-in module for performing MSA, but it provides command-line wrappers for the previously mentioned types of software - ClustalW and MUSCLE.

In addition, there is a module in Biopython for reading alignment files, which are the output of the above-stated types of software. As an illustration, we will use Clustal Omega (https://www.ebi.ac.uk/Tools/msa/clustalo/), an online interface for doing MSA, download the result, and then use Biopython on its output. Before we proceed to MSA, we will first require some sequences for performing MSA. The sequences are given below and are also saved as "multiple_alingment.fasta" as a supplementary file Figure 3.4, Figure 3.5.

```
>Seq0
AACCGTCAGAAGAGAGCTTTATGCAC
>Seq1
ATCCGTCAGGATAGAGTTTTATTCTC
>Seq2
AACCGTCCGAATCGAGCATTATTCTC
>Seq3
AACCGTCCGAATAGCCCTTTATTCTC
>Seq4
ATCGGTACGAATAGAGCTTTATTCTC
>Seq5
AACCGTCGGAAAAGAGCTTTAGTCTC.
```

Open https://www.ebi.ac.uk/Tools/msa/clustalo/.

The Figure 3.4 is the query page of Clustal Omega, where we can paste our sequences for performing MSA. We can also upload our sequences as FASTA files, select the type of sequence and output parameters, and click "Submit" (Figure 3.5).

Once the query has been submitted, we will then obtain a result page, as shown in Figure 3.6. We can download this alignment file using the option highlighted in the figure 3.6 and save it as "aln_file.-clustal" into the Jupyter Notebook directory for further analysis. Biopython has a module called "AlignIO" which can parse the downloaded alignment file in ".clustal" format. Readers are encouraged to open the file in any text editor to see the file format. An example of parsing an alignment file is further illustrated below Figure 3.7):

Code:

```
#open the alignment file
from Bio import AlignIO
with open("aln_file.clustal", "r") as aln:
#use AlignIO to read the alignment file in 'clustal' format
alignment = AlignIO.read(aln, "clustal")
print(alignment)
alignment
```

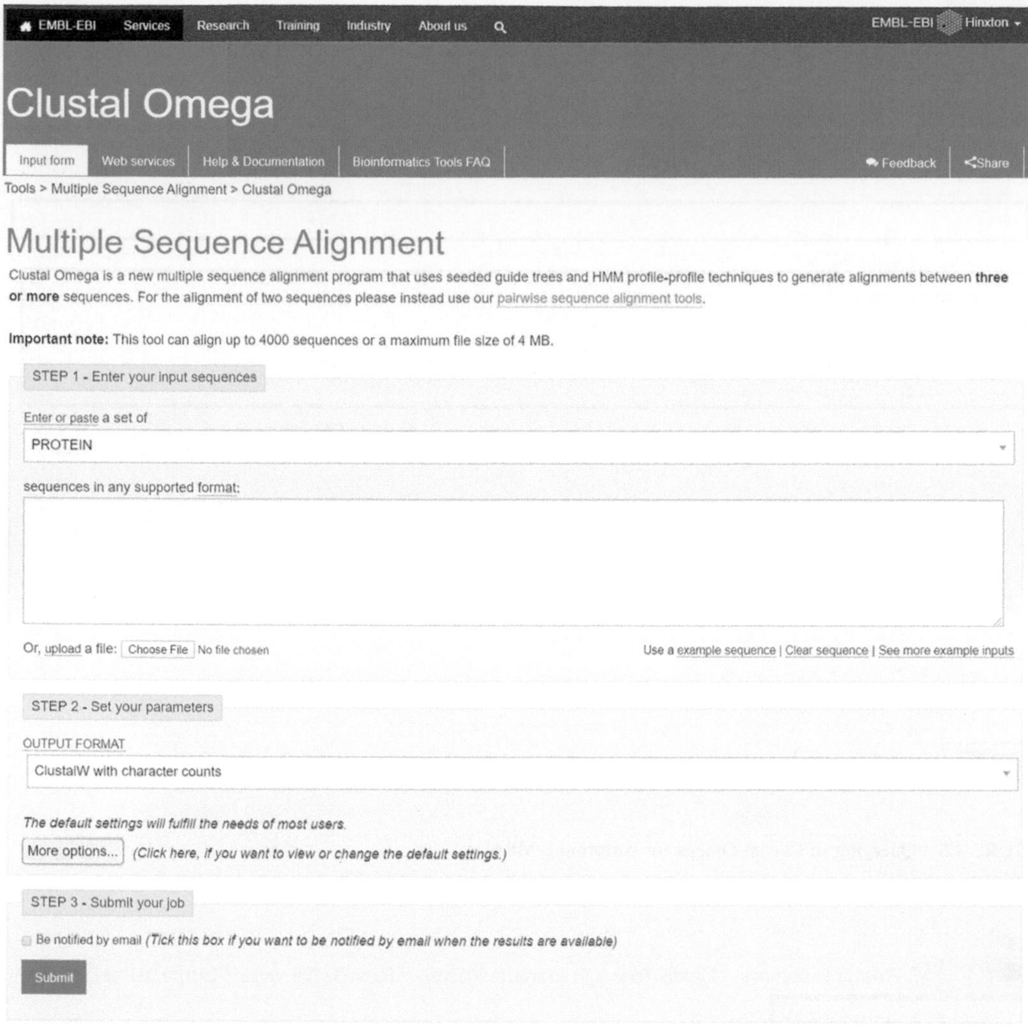

FIGURE 3.4   Query page of Clustal Omega

Output:

```
SingleLetterAlphabet() alignment with 6 rows and 26 columns
ATCCGTCAGGATAGAGTTTTATTCTC Seq1
ATCGGTACGAATAGAGCTTTATTCTC Seq4
AACCGTCAGAAGAGAGCTTTATGCAC Seq0
AACCGTCCGAATAGCCCTTTATTCTC Seq3
AACCGTCCGAATCGAGCATTATTCTC Seq2
AACCGTCGGAAAAGAGCTTTAGTCTC Seq5
```

The output presents a description of the number of sequences as the number of rows and sequence length as a column count. "Bio.AlingIO" can parse several alignment formats; a list of all supported formats is available on https://biopython.org/wiki/AlignIO. By now, we can successfully execute a Pairwise Sequence Alignment. Next, we will study multiple sequence alignment and learn how to plot a phylogenetic tree.

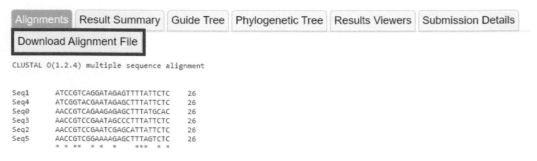

Multiple Sequence Alignment

Clustal Omega is a new multiple sequence alignment program that uses seeded guide trees and HMM profile-profile techniques to generate alignments between **three or more** sequences. For the alignment of two sequences please instead use our pairwise sequence alignment tools.

Important note: This tool can align up to 4000 sequences or a maximum file size of 4 MB.

STEP 1 - Enter your input sequences

Enter or paste a set of

DNA

sequences in any supported format:

>Seq2
AACCGTCCGAATCGAGCATTATTCTC
>Seq3
AACCGTCCGAATAGCCCTTTATTCTC
>Seq4
ATCGGTACGAATAGAGCTTTATTCTC
>Seq5
AACCGTCGGAAAAGAGCTTTAGTCTC

Or, upload a file: Choose File   No file chosen   Use a example sequence | Clear sequence | See more example inputs

STEP 2 - Set your parameters

OUTPUT FORMAT

ClustalW with character counts

The default settings will fulfill the needs of most users.

More options...   *(Click here, if you want to view or change the default settings.)*

STEP 3 - Submit your job

☐ Be notified by email *(Tick this box if you want to be notified by email when the results are available)*

Submit

FIGURE 3.5   Querying in Clustal Omega for performing MSA.

| Alignments | Result Summary | Guide Tree | Phylogenetic Tree | Results Viewers | Submission Details |

Download Alignment File

CLUSTAL O(1.2.4) multiple sequence alignment

```
Seq1    ATCCGTCAGGATAGAGTTTTATTCTC    26
Seq4    ATCGGTACGAATAGAGCTTTATTCTC    26
Seq0    AACCGTCAGAAGAGAGCTTTATGCAC    26
Seq3    AACCGTCCGAATAGCCCTTTATTCTC    26
Seq2    AACCGTCCGAATCGAGCATTATTCTC    26
Seq5    AACCGTCGGAAAAGAGCTTTAGTCTC    26
        *  *  **   *  *   *     ***   *  *
```

FIGURE 3.6   Result of MSA obtained with Clustal Omega

Construction of a Phylogenetic Tree

The construction of phylogenetic trees is one of the widespread applications of MSA. There are several methods of phylogenetic tree construction. One of the most popular ones is the distance-based methods that rely on a measure of "genetic distance" between the sequences that are being classified The sequences having less genetic or distance are more likely to be evolutionary similar, and, hence, will be placed closer in a phylogenetic tree. Therefore, first, we have to calculate the distances among all of the pairs of sequences in our alignment file - thereby, creating a distance matrix.

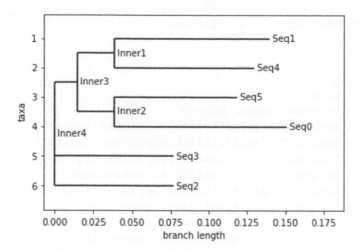

FIGURE 3.7 Output Phylogenetic Tree.

Code:

```
from Bio.Phylo.TreeConstruction import DistanceCalculator
#calculate the distance matrix
calculator = DistanceCalculator('identity')
#adds distance matrix to the calculator object and returns it
dm = calculator.get_distance(alignment)
print(dm)
```

Output:

```
Seq1 0
Seq4 0.192 0
Seq0 0.230 0.269 0
Seq3 0.230 0.192 0.230 0
Seq2 0.230 0.192 0.230 0.153 0
Seq5 0.230 0.230 0.192 0.192 0.192 0
Seq1 Seq4 Seq0 Seq3 Seq2 Seq5
```

In the output, we can determine the pairwise distance in the matrix form that has been calculated using the "alignment" variable from the previous example of parsing an alignment file. Biopython has a module known as "Phylo" that deals with the operations for the construction and plotting of phylogenetic trees. The "DistanceCalculator" object is initiated with a substitution matrix as an argument. An identity matrix is a simple matrix where a score is given for an identity, and "0" is given for mismatch. There will be a slight variation between the output in the workbook and here, as we have rounded the distances up to three decimal places to fit it in the space allotted. Once we have the distances, we can then proceed with constructing a phylogenetic tree, as shown below:

Code:

```
from Bio.Phylo.TreeConstruction import DistanceTreeConstructor
#initialize a DistanceTreeConstructor object based on our distance calcu-
lator object
constructor = DistanceTreeConstructor(calculator)
#build the tree
```

```
tree = constructor.build_tree(alignment)
print(tree)
```

Output:

```
Tree(rooted=False)
Clade(branch_length=0, name='Inner4')
Clade(branch_length=0.014423076923076886, name='Inner3')
Clade(branch_length=0.024038461538461536, name='Inner1')
Clade(branch_length=0.10096153846153844, name='Seq1')
Clade(branch_length=0.09134615384615385, name='Seq4')
Clade(branch_length=0.024038461538461536, name='Inner2')
Clade(branch_length=0.0801282051282051, name='Seq5')
Clade(branch_length=0.1121794871794872, name='Seq0')
Clade(branch_length=0.07692307692307694, name='Seq3')
Clade(branch_length=0.07692307692307693, name='Seq2')
```

Biopython's "Phylo.TreeConstruction" has a module called "DistanceTreeConstructor". It identifies the "DistanceCalculator" object as an argument, which contains the substitution matrix for calculating the distances. Next, we can place our alignment variable to the tree constructer for building a phylogenetic tree. The output may take some effort to understand, where each line represents a node and tabs are branch lengths. To make it simple, we can also draw/plot this tree (Figure 3.7):

Code:

```
from Bio import Phylo
import pylab
#draw the tree
Phylo.draw(tree)
```

Output:

When drawing a phylogenetic tree, we will require a "pylab" module. This comes with Anaconda by default, or it can be installed using the "pip" or "conda" installers. The output above is easy to interpret for common ancestors, speciation points, etc. Now, the previous example output can also make some sense according to the above figure.

Handling PDB Files

The Protein Data Bank (PDB) is a repository of 3D protein structures. All biologists would know that the 3D folding of the amino acid chain determines its function. 3D structures can be experimentally determined by X-ray crystallography, NMR, and electron microscopy. Among these, X-ray crystallography is the most widely used. It is so common that more than 80% of the structures in the PDB database are determined by X-ray crystallography. The experimental identification of the 3D structure is essentially a deduction of the coordinates of each atom in a 3-dimensional space (i.e. x, y, and z coordinates). These coordinates, along with some annotations, are stored in a particular file format called the ".pdb" format. The structure of the PDB format is shown below. Each line representing the coordinates of the atoms is divided into several columns:

```
HEADER OXYGEN TRANSPORT 22-MAR-79 1MBS
TITLE X-RAY CRYSTALLOGRAPHIC STUDIES OF SEAL MYOGLOBIN. THE
TITLE 2 MOLECULE AT 2.5 ANGSTROMS RESOLUTION
```

```
COMPND MOL_ID: 1;
COMPND 2 MOLECULE: MYOGLOBIN;
...
EXPDTA X-RAY DIFFRACTION
AUTHOR H.SCOULOUDI
...
REMARK 350 BIOMOLECULE: 1
REMARK 350 AUTHOR DETERMINED BIOLOGICAL UNIT: MONOMERIC
REMARK 350 APPLY THE FOLLOWING TO CHAINS: A
REMARK 350 BIOMT1 1 1.000000 0.000000 0.000000 0.00000
REMARK 350 BIOMT2 1 0.000000 1.000000 0.000000 0.00000
REMARK 350 BIOMT3 1 0.000000 0.000000 1.000000 0.00000
...
SEQRES 1 A 153 GLY LEU SER ASP GLY GLU TRP HIS LEU VAL LEU ASN VAL
SEQRES 2 A 153 TRP GLY LYS VAL GLU THR ASP LEU ALA GLY HIS GLY GLN
SEQRES 3 A 153 GLU VAL LEU ILE ARG LEU PHE LYS SER HIS PRO GLU THR
...
ATOM 1 N GLY A 1 15.740 11.178 -11.733 1.00 0.00 N
ATOM 2 CA GLY A 1 15.234 10.462 -10.556 1.00 0.00 C
ATOM 3 C GLY A 1 16.284 9.483 -9.998 1.00 0.00 C
ATOM 4 O GLY A 1 17.150 8.979 -10.709 1.00 0.00 O
ATOM 5 N LEU A 2 16.122 9.240 -8.705 1.00 0.00 N
...
HETATM 1225 CHA HEM A 154 9.596 -13.100 10.368 1.00 0.00 C
HETATM 1226 CHB HEM A 154 11.541 -10.200 7.336 1.00 0.00 C
HETATM 1227 CHC HEM A 154 9.504 -6.500 9.390 1.00 0.00 C
...
```

1. The first column contains the type - such as atoms or heteroatoms atoms - where atoms consist of elements found in general proteins, nucleic acids, and/or heteroatoms.
2. The next column is the serial number of the atom.
3. The third column is the atom itself (i.e. alfa, beta, gamma, etc), according to its position in a residue.
4. Residues are presented in the fourth column. The chain where the residue is present and the serial number of the residue occupies the next two columns, respectively.
5. The next three columns are the x, y, and z coordinates of the atom in a 0.1-nanometer scale. Following this, the two columns are the occupancy and temperature factor - which show the confidence of the coordinates. The last column shows the element itself.

Figure 3.8 is the 3D visualization of the above-stated 3D file along with ligand heteroatoms in the RCSB PDB website (https://www.rcsb.org/structure/1MBS):

Various software like RasMol, PyMOL, JMOL, Chimera, etc. are used for 3D visualization of proteins using PDB files. Biopython has a module dedicated to the handling of 3D structure files, and it can be imported by "Bio.PDB".

Let us practice some of the functions of "Bio.PDB". First, we will download a 3D structure file in the "pdb" and "mmCif" format.

Hands on Data Science for Biologists

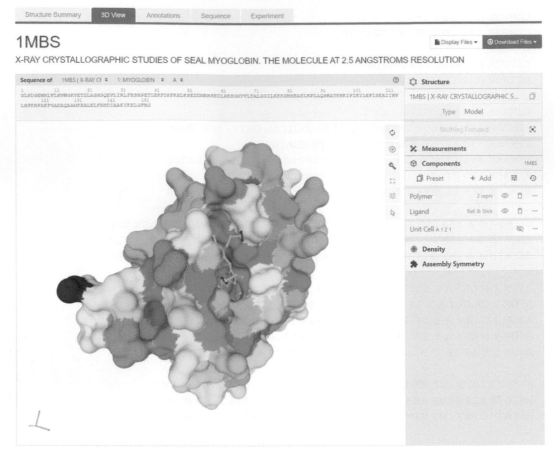

FIGURE 3.8   3D Structure Snapshot of X-ray crystallographic study of SEAL Myoglobin, PDB ID: 1MBS

Code:

```
from Bio.PDB import PDBList
pdbl = PDBList()
pdbl.retrieve_pdb_file('1mbs', pdir = '.', file_format = 'pdb')
pdbl.retrieve_pdb_file('1mbs', pdir = '.', file_format = 'mmCif')
```

Executing the above commands will download two files in the directory. The "retrive_pdb_file" is a function inside the "PDBList" which takes three arguments, PDB ID, directory, and format of the file, and save the file in the desired directory. Here, we configure PDB ID as "1mbs", directory as "." (i.e. implying the current directory), and formats as "pdb" and "mmCif". The PDBx or mmCIF format which uses data dictionaries to define the information content of PDB entries is used. This format has a flexible key-value pair to represent structural macromolecular information and can accommodate any number of atoms, residues, or chains. "Bio.PDB" has a module called "PDBParser", which can read PDB files and can save pieces of information in a structured format such as a Python dictionary so that data can be retrieved easily.

```
Code:

from Bio.PDB.PDBParser import PDBParser
parser = PDBParser(QUIET = True)
```

```
pdb_id = "1mbs"
pdb_filename = "pdb1mbs.ent"
structure_pdb = parser.get_structure(pdb_id, pdb_filename)
print(structure_pdb.header.keys())
print('----------------------------')
print(structure_pdb.header["name"])
print(structure_pdb.header["release_date"])
print(structure_pdb.header["resolution"])
print(structure_pdb.header["keywords"])
```

Output:

```
dict_keys(['name', 'head', 'idcode', 'deposition_date',
'release_date', 'structure_method', 'resolution',
'structure_reference', 'journal_reference', 'author', 'compound',
'source', 'has_missing_residues', 'missing_residues', 'keywords',
'journal'])


----------------------------
x-ray crystallographic studies of seal myoglobin. the molecule at 2.5 ang-
stroms resolution
1979-05-15
2.5
oxygen transport
```

Here, "pdb1mbs.ent" is the PDB format structure file that we had downloaded in the previous example and saved in the current directory. We have initiated "PDBParser" with the argument "QUITE == True". This will ignore any errors with the PDB file while parsing it - such as missing residues, information, etc. Therefore, we have an object "structure_pdb" after having parsed information of the PDB file. The header information can be accessed as "structure_pdb.header()" which contains information in Python dictionary format, as shown in the example above. Data - such as name, release date, and resolution - can be retrieved via respective keys. We can also access each of the models, chains, residues, and atoms as objects using the "structure_pdb" object. These objects can be accessed in a particular hierarchy. On top, we have models. A PDB file may contain more than one model.

Code:

```
models = list(structure_pdb.get_models())
print(models)
```

Output:

```
[<Model id=0>]
```

".get_models()" from the structure object will list all of the models. In this case, we have only one model. Next, we shall find the chains in the model.

Code:

```
chains = list(models[0].get_chains())
print(chains)
```

Output:

```
[<Chain id=A>]
```

The 1MBS file contains only one chain - chain "A". In the future, we can retrieve residues of this chain using the ".get_residues" function of the chain object.

Code:

```
residues_0 = list(chains[0].get_residues())
print(len(residues_0))
```

Output:

```
154
```

The chain "A" of 1MBS contains 154 residues. In the next example, we will grab all of the atoms of the first residue as well as the coordinates of the first model.

Code:

```
print(residues_0[0])
atoms = list(residues_0[0].get_atoms())
print(atoms)
print(atoms[0].get_vector())
```

Output:

```
<Residue GLY het= resseq=1 icode= >
[<Atom N>, <Atom CA>, <Atom C>, <Atom O>]
<Vector 15.74, 11.18, -11.73>
```

The output shows that the first residue of chain "A" is glycine, and it contains four atoms. The x, y, and z coordinates of the first atom are 15.74, 11.18, and -11.73 respectively. To summarize, the hierarchy is as follows - models, chains, residues, and lastly, the atoms. Let us combine all of the topics together to find the coordinates of all of the atoms of the first residue of the 1-MBS PDB file (i.e. glycine).

Code:

```
for model in structure_pdb.get_models():
    for chain in model.get_chains():
        for residue in chain.get_residues():
            for atom in residue.get_atoms():
            print(atom)
            print(atom.get_vector())
            break
            break
    break
```

Output:

```
<Atom N>
<Vector 15.74, 11.18, -11.73>
<Atom CA>
<Vector 15.23, 10.46, -10.56>
<Atom C>
<Vector 16.28, 9.48, -10.00>
<Atom O>
<Vector 17.15, 8.98, -10.71>
```

Breaks are used to limit the process by up to one residue. This way, we can print any atom and its corresponding coordinate.

The other 3D macromolecular file format "mmCif2" is relatively new and supported with every protein 3D structure database. The "mmCif2" format is flexible, as it is in the dictionary type file structure. Any information can be saved using new keys. We have a download file of 1MBS in the "mmCif2" format. Let us extract information from it below:

Code:

```
from Bio.PDB.MMCIF2Dict import MMCIF2Dict
mmcif_dict = MMCIF2Dict("1mbs.cif")
print(list(mmcif_dict.keys())[:10])
print(len(list(mmcif_dict.keys())))
```

Output:

```
['data_', '_entry.id', '_audit_conform.dict_name', '_audit_conform.dict_
version', '_audit_conform.dict_location', '_database_2.database_id', '_da-
tabase_2.database_code', '_pdbx_database_status.status_code', '_pdbx_data-
base_status.entry_id',
'_pdbx_database_status.recvd_initial_deposition_date']
 528
```

The "MMCIF2Dict" reads a ".cif" file and converts it into the Python dictionay. We have printed ten keys, and the number of keys were discovered to be 528. Next, we will explore some more keys.

Code:

```
print(mmcif_dict["_exptl_crystal.density_percent_sol"])
y_list = mmcif_dict["_atom_site.Cartn_y"]
print(len(y_list))
print(y_list[:10])
```

Output:

```
['52.12']
1266
['11.178', '10.462', '9.483', '8.979', '9.240', '8.164', '7.481', '7.424',
'7.101', '7.724']
```

The "_exptl_crystal.density_percent_sol" key gives the density of the crystal, and "_atom_site.Cartn_y" includes all of the y coordinates of the 3D structure. There are a total of 1,266 y coordinates, meaning 1,266 atoms are present in 1MBS structure.

There are various ways in which we can analyze 3D structures, including:

1. Distance measurement
2. Angle measurement
3. Torsional angles calculation
4. Atom-atom interactions certainty
5. Superimposing structures
6. Residues mapping on each other of two related structures
7. Identification of the secondary structure
8. Residue size estimation

We will observe the implementation of distance measurement between two atoms in 3D space. The distance between a point P1(x1,y1,z1) and p2(x2,y2,z2) can be calculated as the Euclidean distance.

Code:

```
import math
coordinates_0 = atoms[0].get_vector()
coordinates_1 = atoms[-1].get_vector()
x0,y0,z0 = coordinates_0
x1,y1,z1 = coordinates_1
#Euclidean distance
distance = math.sqrt((x1 - x0)**2+(y1 - y0)**2+(z1 - z0)**2)
print(round(distance,6))
```

Output:

```
2.805758
```

We have gathered the first and last atoms from the previous example of the first residue. The distance is 2.8057 Å(Armstrong). There is a more simple way to find the distance between two atom objects of Biopython, and this is done by just subtracting them.

Code:

```
atoms = list(residues_0[0].get_atoms())
print(atoms[0]-atoms[-1]) #builtin in pdb module
```

Output:

```
2.8057582
```

This output corroborates with the previously obtained output.

In this chapter, we have discussed some important modules of Biopython along with their usage. In the next chapter, we will study the excel of Python, as well as an efficient data analysis module, and do some analysis using real-world data. Let us continue this journey together.

Exercise

1. Create a DNA "seq" object using the sequence "ATCCGGAATCCG".
2. With the above "seq" object, find its reverse complement and its translated protein sequence.
3. Read the FASTA file "exercise.fasta" provided in the supplementary material and print the sequences.
4. The sequences above are 16S mRNAs of certain microorganisms. Perform BLAST to identify the species.
5. Apply multiple sequence alignment on the sequences and construct a phylogenetic tree using BioPython Packages.

4

Python for Data Analysis

Introduction

In data analysis pipelines, data is commonly structured in table formats - where columns represent features, properties, or characteristics of data, and rows serve as the instances or observations of those characteristics. The management of these rows and columns is one of the necessary skills of a data scientist. These rows and columns constitute a matrix with arrays of data. In chapter 2 we have learned about the concept of "nested lists" (recall Figure 2.4), where we defined a nested list as a 2D table or matrix. We have also learned about and discussed a standard table format file ".csv", where columns in the data are separated by commas, and the data could be loaded as a nested list using Python's built-in library "CSV".

Loading data in the "list" format is always preferable, but processing data in this format is not always advisable in data analysis due to a variety of reasons, such as:

1. Manipulating the data in a list requires "for" loops, and these are considered slow for large lists.
2. Pieces of data are mostly in numerical form (i.e. number type), and we know that lists are designed to be used for any datatype. The list is, therefore, not optimized for number operations.
3. Finally, the "list" format is not memory efficient.

To overcome these lacunae in Python's built-in data structure, a package called "NumPy", an abbreviation for Numerical Python, has been developed. This library is very effective in handling numbers and matrices. NumPy also contains various functions for matrix operations, linear algebra, statistics, and many more.

Data in row and column formats can also be represented in spreadsheets for analysis. Spreadsheets are important for storing and organizing data or information. These are used for preliminary data analysis such as rearranging data, applying the formula or functions to data, and modifying data, among others. Some examples of popular spreadsheet types of software include Microsoft Office Excel, Libra Office Spreadsheets, and Google Spreadsheets. There is a package called "Pandas" in Python, which can handle data like spreadsheets and aid in preliminary data analysis and visualization of data. The name has been derived from "panel data", and it is an econometric word that is utilized for multidimensional structured data. Spreadsheets are very useful for handling any row-column structured data and are most favored in business or economic data handling.

NumPy

NumPy Arrays versus Lists

We will introduce NumPy by comparing it with "lists" so that its application is already familiar to the readers. The first instruction is to import NumPy and create an array, as explained below:

Code:

```
import numpy as np
l = [1,2,3]
```

```
a = np.array([1,2,3])
print(l)
print(a)
```
Output:
```
[1, 2, 3]
[1 2 3]
```

In this example, we imported the NumPy package and attached the "as" keyword with np. We can import packages with the "as" keyword, which helps us to rename the package in our script - preferably in shortening the name of the package so that we do not have to type the full name repeatedly. NumPy has been abbreviated as np, which is a standard convention for importing NumPy into a script. However, it is not mandatory to import it as "np". We can write anything other than "np" and use it in our script instead of typing "numpy". In the second line, we created a list containing the numbers [1, 2, 3]. Subsequently, we created a NumPy array using the "np.array()" function, which uses a list as an argument and returns a NumPy array. Printing both the "list" and the "NumPy" array shows a small difference (i.e. "list" items are separated by commas, which is not the case for "NumPy" arrays).

We can iterate over items of "NumPy" and "List" using a "for" loop.

Code:
```
for i in l:
    print(i)
print('---------')
for i in a:
    print(i)
```
Output:
```
1
2
3
---------
1
2
3
```

This operation is the same for lists and arrays. Next, we will do some operations such as adding and multiplying values in the NumPy array, as illustrated below:

Code:
```
print(l + [4])
print(a + np.array(4))
print('------')
print(2*a)
print(2*l)
```
Output:
```
[1, 2, 3, 4]
[5 6 7]
------
[2 4 6]
[1, 2, 3, 1, 2, 3]
```

In the code above, we added [4] to list and array(4) to our NumPy array using "l" and "a" variables from the previous code for the list and the array, respectively. In the output, we can observe how differently lists and arrays deal with addition. The list adds a new item in the previous list, whereas the array does element-wise addition (i.e. it adds four to every element of the array). The same is apparent in the case of multiplication. Therefore, there exists a point of caution; mathematical operations are different for Python lists and NumPy arrays. In the next example, we will study how to use power notation for each element of a list and an array:

Code:
```
l2 = []
for e in l:
    l2.append(e**2)
print(l2)
print(a**2)
```
Output:
```
[1, 4, 9]
[1 4 9]
```
In order to square every element of a list, we first have to create a new and empty list, use a "for" loop to square every element of a list, and, lastly, add these to the new list. In the case of arrays, we can operate "**" to every element of the array in one line - for example, "a**2" will square all of the elements of the array, which is comparatively faster than applying a "for" loop and has a lesser number of commands.

Let us examine a few more built-in functions of arrays:

Code:
```
print(np.sqrt(a)) # Square Root
print(np.log(a)) # Finding Log
print(np.exp(a)) # Exponent
print(np.tanh(a)) # hyperbolic tangent
```
Output:
```
[1. 1.41421356 1.73205081]
[0. 0.69314718 1.09861229]
[ 2.71828183 7.3890561 20.08553692]
[0.76159416 0.96402758 0.99505475]
```
These are some of the built-in functions in NumPy. There are many similar functions that readers can find in NumPy documentation. NumPy operates element-wise, because we have to perform operations on columns or rows most of the time.

Two-Dimensional Matrices

NumPy has been used for operating on higher-dimensional data. In the case of lists, the nested list is two-dimensional data. Let us figure out what the case is for NumPy arrays in the succeeding example:

Code:
```
import numpy as np
l = [[1,2,3],[4,5,6]]
print(l)
n = np.array(l)
print(n)
```
Output:
```
[[1, 2, 3], [4, 5, 6]]
[[1 2 3]
 [4 5 6]]
```

TABLE 4.1

2D NumPy Array

1	2	3
4	5	6

In this example, the "l" variable contains a nested list. When "l" is passed to "np.array()" as an argument, it is then converted into a 2D matrix. Printing a nested list and an array shows us the difference. Next, we will look for the index of elements in a 2D array and how to access an item from a particular row, column, or both (Table 4.1).

Code:
```
#Selecting the number from the first row of the second column
print(l[0][1])
print(n[0][1])
print(n[0,1])
# Selecting first column
print(n[:,0])
# Selecting first row
print(n[0,:])
```
Output:
```
2
2
2
[1 4]
[1 2 3]
```

From the example above, we can use the conventional square bracket notation for accessing data inside a NumPy array, and this is the same as in nested lists. For the 2D matrix, the index inside the first bracket identifies the row, and the second index represents the column. NumPy arrays also have single square bracket notation which takes two indexes separated by a comma, like:

"n[0,1])"where the first element refers to the row number, and the second element refers to the column number. All Python indexes start from "0" instead of "1", and this stands true for NumPy arrays also.

Inside the single bracket convention, we can also use a colon for selecting all of the items between indexes of a particular row and column in respective places, similar to lists and strings as seen in the fourth and fifth statements in the code block above.

In the code above, we have shown how to choose the first row and first column, which the user can match with the matrix given in Table 4.1. Readers can create 2D matrices and practice indexing with the Jupyter Notebook provided with this chapter. We can also create 3D, 4D, or up to *n*-dimensional matrices using NumPy, with index notations remaining the same. The reader may give it a try by creating a 3D matrice.

Matrix Operations

NumPy has various applications in matrix operations, linear algebra operations, generating numbers, and many more. Most of the Machine Learning and image processing packages in Python are based on NumPy arrays due to their lucid handling of numbers and matrices. In this section, we shall describe the functionalities of NumPy built-in libraries as a mathematical package.

Code:
```
n = np.array([[1,2,3],[4,5,6]])
# transpose of a matrix
print(n.T)
print('---------')
# natural exponential function
print(np.exp(n))
```
Output:
```
[[1 4]
 [2 5]
 [3 6]]
```

```
----------
[[ 2.71828183 7.3890561 20.08553692]
 [ 54.59815003 148.4131591 403.42879349]]
```

One simple operation with a matrix is transposing it or interchanging its row and columns. In NumPy, ".T" returns a transposed matrix. In applying a natural exponential function, NumPy has a built-in ".exp ()" function. This function considers an array as an argument, but we can also pass it a list as its arguments. NumPy will first convert it into an array and then return the values, as shown in the code below:

Code:

```
print(np.exp([[1,2,3],[4,5,6]])) #List treated as numpy array
```

Output:

```
[[ 2.71828183 7.3890561 20.08553692]
 [ 54.59815003 148.4131591 403.42879349]
```

Matrix multiplication or the dot product is the most popular and essential of matrix operations. It is widely used in Machine Learning, deep learning, solving linear algebra systems, population modeling, and network biology, to name a few.

The dimension of matrices is defined by the number of rows X the number of columns. For example, in the above figure matrix, 'M' has a 3×2 dimension. Figure 4.1 shows how the multiplication of matrices occurs. For example, matrix 'M' has a 3×2 dimension, and matrix 'N' has a 2×1 dimension. Therefore, the resultant matrix '$M \times N$' will have a 3×1 dimension. It should be noted that, for multiplication of two matrices, the inner dimensions should be equal or the number of columns of one matrix should be equal to the number of rows of another matrix. The resultant matrix dimensions will be equal to the outer dimensions, and, in this case, it is 3×1. Let us perform matrix multiplication in NumPy using the example below:

Code:

```
import numpy as np
a = np.array([[1,2,3],[4,5,6]]) #Shape(2×3)
b = np.array([[1,2],[3,4],[5,6]])#Shape(3×2)
print(a.dot(b)) # dot product
```

Output:

```
[[22 28]
 [49 64]]
```

Matrix multiplication is also called the dot product. In this example, we created two matrices with dimensions 2×3 and 3×2 and used an instance method of NumPy arrays ".dot()" which considers an array as an argument. The resultant matrix has a dimension of 2×2. What if we attempt to do multiplication of two matrices whose inner dimensions are not the same? It is important to mention that the "*" operation on NumPy arrays carries out element-wise multiplication and not dot product or matrix multiplication.

Code:

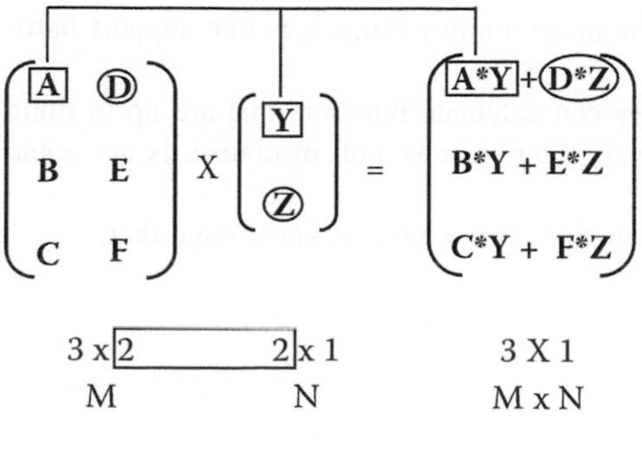

FIGURE 4.1 Matrices Multiplication Operation (Rows *X* Columns).

```
import numpy as np
a = np.array([[1,2,3],[4,5,6]]) #Shape(2×3)
b = np.array([[1,2],[3,4]])#Shape(2×2)
print(a.dot(b)) # dot product
```
Output:

```
ValueError
<ipython-input-35-d1b890f4fe85> in <module>
      2 a = np.array([[1,2,3],[4,5,6]]) #Shape(3×2)
      3 b = np.array([[1,2],[3,4]])#Shape(2×2)
----> 4 print(a.dot(b)) # dot product
ValueError: shapes (2,3) and (2,2) not aligned: 3(dim 1) != 2(dim 0)
```
Yes, we got the result we expected - an error stating that the dimensions are misaligned.

Finding the determinant and determining the inverse of matrices are crucial operations of linear algebra. NumPy has a library named "linalg" to deal with these operations of linear algebra. In NumPy, we can discover these determinants and the inverse by using a single command:

Code:
```
import numpy as np
a = np.array([[5,6],[7,8]]) #Shape(2×2)
# Matrice determinant
print(np.linalg.det(a))
# Matrice inverse
print(np.linalg.inv(a))
```
Output:
```
-1.999999999999999
[[-4. 3.]
 [ 3.5 -2.5]]
```
"np.lialg" has "det()" and "inv()" methods that take a matrix and return its determinant and inverse, respectively.

We can also confirm if the inverse is true or not, because multiplying a matrix with its inverse produces an identity matrix - which has ones in the diagonal and zeros everywhere else. Let us multiply our matrix with its inverse below:

Code:
```
#Checking the inverse
a = np.array([[5,6],[7,8]])
print(np.linalg.inv(a).dot(a))
```
Output:
```
[[ 1.00000000e+00 0.00000000e+00]
 [-3.55271368e-15 1.00000000e+00]]
```
We obtained a matrix that is very near to the identity matrix. Note that this is a bug that is neither in Python nor in the user's code. This is true for most of the programming languages that support hardware's floating-point arithmetic.

We know that our computers have finite memory. They can calculate numbers that are up to finite precisions. $-3.552e-15$ means $-3.552*10-15$, which is very near to zero, but, of course, is not equal to zero.

We can use the inverse of a matrix to solve a linear equation. It is common knowledge that:

$$Ax = B$$

$$x = BA^{-1}$$

We can solve an equation in the linear system by multiplying the inverse of the transformation matrix with the product matrix.

Let us solve for the given equations:

$$-3x - 2y + 4z = 9$$

$$3y - 2z = 5$$

$$4x - 3y + 2z = 7$$

Code:
```
## Solving linear equations
a = np.array([[-3,-2,4],[0,3,-2],[4,-3,2]])
b = np.array([9,5,7])
print(np.linalg.inv(a).dot(b))
```
Output:
```
[3. 7. 8.]
```
Therefore, we received the output $x = 3$, $y = 7$, and $z = 8$. We can make this more efficient and straightforward.

The "linalg" library contains a method called "solve", which can be used for solving linear equations. It is faster than the method we previously used above and is recommended for solving matrices in NumPy.

Code:
```
print(np.linalg.solve(a,b))
```
Output:
```
[3. 7. 8.]
```

Comparing Matrices

When comparing matrices, there are two types of comparisons. One is the element-wise comparison, and the other is a full matrix comparison.

Code:
```
## Camparing matrice
a = np.array([[1,2],[4,6]]) #Shape(2×2)
b = np.array([[1,2],[3,4]]) #Shape(2×2)
c = np.array([[1,2],[4,6]]) #Shape(2×2)
print(a==b)
print(a==c)
print('-------------')
print(np.allclose(a,b))
print(np.allclose(a,c))
```
Output:
```
[[ True True]
 [False False]]
[[ True True]
 [ True True]]
-------------
False
True
```
In the above example, we have created three matrices for the sake of a demo of matrix comparisons in NumPy. Simple "==" will make an element-wise comparison and expresses "True" in the place where elements of both matrices are equal and "False" in the place where elements of both matrices are

unequal. We know that True and False are not strings for Python; rather, they are bool datatype. To compare all of the matrices at a time with a single bool value, we can use a built-in function "np.allclose ()". The "np.allclose()" not only compares two matrices, but it also takes into account the value precision problem which we discussed above in the case of identity matrices. A straightforward comparison will return False for "0= =−3.552e−15", but the allclose() method will return it as True.

Generating Data Using NumPy

Generating synthetic data is often required for testing the robustness of Machine Learning algorithms and other computer software. We have discussed Python's built-in library "random" in chapter 2, which can be used to create random numbers or choose items randomly from a list. NumPy also can generate high-dimensional data. First, we will use NumPy to create matrices of ones, zeros, or any number with user-defined dimensions, as seen below:

Code:
```
print(np.zeros((2,3)))
print('------------')
print(np.ones((2,3)))
print('------------')
print(10*np.ones((2,3)))
print('------------')
print(np.eye(3))# Identity matrices
```
Output:
```
[[0. 0. 0.]
 [0. 0. 0.]]
------------
[[1. 1. 1.]
 [1. 1. 1.]]
------------
[[10. 10. 10.]
 [10. 10. 10.]]
------------
[[1. 0. 0.]
 [0. 1. 0.]
 [0. 0. 1.]]
```
Methods such as "np.zeros()" and "np.ones()" take dimension as a tuple in argument and produce a matrix of 0's or 1's with user-defined dimensions. The generated arrays of ones can be multiplied with a number to generate a matrix of that number - like 10 - as shown in the example above. "np.eye()" produces identity matrices with the dimensions argument. NumPy can also generate random numbers, random numbers following a special probability, or geometric distributions.

Code:
```
print(np.random.random())
print('------------')
print(np.random.random((2,3)))
print('------------')
print(np.random.randn(2,3))
```
Output:
```
0.7420709390526112
------------
[[0.64231956 0.27470495 0.83306675]
 [0.62637418 0.870846 0.10071372]]
------------
```

```
[[-0.85548405 0.94454349 -0.09391225]
 [ 0.27754079 -0.92784648 -0.56905913]]
```

"np.random.ramdom()" can generate a random number, as well as a matrix of random numbers with user-defined dimensions. "np.random.randn()" generates a matrix of random numbers forming a Gaussian or uniform distribution.

Additionally, there are methods in NumPy for calculating simple statistics.

Code:

```
r = np.random.randn(10000)
print(r.mean())
print(r.var())
print(r.std())
```

Output:

```
0.0019385521899210887
0.9913490868162517
0.9956651479369214
```

'.mean()', '.var()', and ".std()" return mean, variance, and standard deviation of data respectively, which can present an idea about the development of data. We can see that the mean and standard deviation of this generated data are near 0 and 1, respectively. This clearly indicates that the generated data follows a normal distribution.

Speed Test

In this chapter, we have emphasized and we continue to affirm that the operations of NumPy are much faster than using "for" loops in the nested list. Next, we will do a speed test for calculating the dot product of two randomly generated arrays of length 100 by using the "for" loop and NumPy's ".dot()" method. We will then calculate the dot product 100,000 times using both methods and determine the ratio of the time duration.

Code:

```
from datetime import datetime
a = np.random.rand(100)
b = np.random.rand(100)
T = 100000
def slow_dot_product(a,b):
    result = 0
    for e,f in zip(a,b):
        result += e*f
    return result
t0 = datetime.now()
for t in range(T):
    slow_dot_product(a,b)
dt1 = datetime.now() - t0
t0 = datetime.now()
for t in range(T):
    np.dot(a,b)
dt2 = datetime.now() - t0
print("d1 / dt2:",dt1.total_seconds()/
    dt2.total_seconds())
```

Output:

```
d1 / dt2: 34.875979378699014.
```

In this code block, we observe that NumPy's ".dot()" method is 34.87 times faster than the standard Python loop for the above operation. If we increase the length of the arrays to 1,000 and repeat the same speed test, NumPy's ".dot()" turns out to be 300 times faster.

The function "slow_dot_product" calculates the dot product using a Python loop, and the "datetime" package is used for tracking the time.

We shall now move to the next section of this chapter which is about data-handling using Pandas.

"Pandas" Dataframe

The "Pandas" library is used for the tabular representation of data, where we can do primary data analysis, processing, and more. The "Pandas" library is already installed in the Anaconda distribution of Python. If readers are using vanilla Python, then they can install the "Pandas" library using the "pip" package manager. "Pandas" can load data from most of the tabular data formats like CSV, TSV, Excel files, and SQL databases, among others. As an illustration, we will load a dataset called "Pima Indians Diabetes Database" in CSV format, which we have downloaded from Kaggle using this link "https://www.kaggle.com/uciml/pima-indians-diabetes-database". Readers can find the dataset names as "datasets.csv" in the supplementary folder with this chapter. Kaggle is an open community of data scientists where various datasets and processing pipelines are shared. It is also a subsidiary of Google LLC. Kaggle is an excellent learning platform for Machine Learning researchers.

"Pandas" is imported as "pd" and, just as "NumPy" is imported as "np", this is the most common abbreviation. Let us now dive into the Jupyter Notebook and import Pandas.

Code:
```
import pandas as pd
df = pd.read_csv('datasets.csv')
print(type(df))
```
Output:
<class 'pandas.core.frame.DataFrame'>

Pandas' ".read_csv()" method loads a CSV file into the Pandas dataframe. We can confirm the type of "df" variable: it is a dataframe object. We can observe how the data is structured inside a dataframe using the ".head()" method.

Code:
```
df.head()
```
Output:
The output is a table containing five data rows and a top header row (Table 4.2). The "head" function returns the top five rows from the dataframe. We can consider integers as an argument in the ".head()" method to see the corresponding numbers of rows.

Like the ".head()" method, the dataframe also has the ".tail()" method that displays the last five rows of the dataframe.

TABLE 4.2

df.head() Output

	Pregnancies	Glucose	Blood Pressure	Skin Thickness	Insulin	BMI	Diabetes Pedigree Function	Age	Outcome
0	6	148	72	35	0	33.6	0.627	50	1
1	1	85	66	29	0	26.6	0.351	31	0
2	8	183	64	0	0	23.3	0.672	32	1
3	1	89	66	23	94	28.1	0.167	21	0
4	0	137	40	35	168	43.1	2.288	33	1

TABLE 4.3

Output of df.tail()

	Pregnancies	Glucose	Blood Pressure	Skin Thickness	Insulin	BMI	Diabetes Pedigree Function	Age	Outcome
763	10	101	76	48	180	32.9	0.171	63	0
764	2	122	70	27	0	36.8	0.340	27	0
765	5	121	72	23	112	26.2	0.245	30	0
766	1	126	60	0	0	30.1	0.349	47	1
767	1	93	70	31	0	30.4	0.315	23	0

Code:

```
df.tail()
```

Output (Table 4.3):

The first column is the index of the row of the dataframe.

Code:

```
df.info()
```

Output:

```
<class 'pandas.core.frame.DataFrame'>
RangeIndex: 768 entries, 0 to 767
Data columns (total 9 columns):
Pregnancies                768 non-null int64
Glucose                    768 non-null int64
BloodPressure              768 non-null int64
SkinThickness              768 non-null int64
Insulin                    768 non-null int64
BMI                        768 non-null float64
DiabetesPedigreeFunction   768 non-null float64
Age                        768 non-null int64
Outcome                    768 non-null int64
dtypes: float64(2), int64(7)
memory usage: 54.1 KB
```

The ".info()" method provides the overall description of the data content of the dataframe. The imported database - Pima Indians Diabetes Database - contains 768 entries in nine columns. The nine columns include the attributes "Pregnancies", "Glucose", "BloodPressure", "SkinThickness", "Insulin", "BMI", "DiabetesPedigreeFunction", "Age", and "Outcome". All of the other columns, except "Outcome", contain a reading or the features of patients, and the outcome predicts if the patient is likely to be susceptible to diabetes or not in binary format (i.e. "0" for no and "1" for yes). This data can be used to train a Machine Learning algorithm to predict the susceptibility of a certain individual to diabetes by entering the nine attribute values. This part will be covered in chapter 12 of the book.

Now, let us study what other pieces of information the ".info()" method can provide. It demonstrates if the columns contain any null values, as well as illustrates the datatype of the values in the columns. Lastly, it shows memory usage or the memory this dataframe acquires in the RAM.

Selecting Rows and Columns

We can grab all of the column names as a Python list.

Code:

```
df.columns
```

Output:
```
Index(['Pregnancies', 'Glucose', 'BloodPressure', 'SkinThickness',

'Insulin','BMI', 'DiabetesPedigreeFunction', 'Age', 'Outcome'],
    dtype = 'object')
```
The dataframe object contains a list of columns in a variable called "columns". We can identify the list by calling that variable of the dataframe object. We can also change that variable by assigning it to another list containing new column names. Let us substitute some column names with their abbreviations, like "BP" for "BloodPressure", "ST" for "SkinThickness", and "DPF" for "DiabetesPedigreeFunction". This will transform our dataframe into something visually optimal.

Code:
```
df.columns = ['Pregnancies', 'Glucose', 'BP', 'ST', 'Insulin',
        'BMI', 'DPF', 'Age', 'Outcome']
df.head()
```
Output (Table 4.4):

In this table, we can regard that, by passing a list of new column names to the columns variables, we have changed the column names. Next, we will learn how to select columns.

Code:
```
#Selecting Columns
print(df['Glucose'])
print(type(df['Glucose']))
```
Output:
```
0     148
1      85
2     183
3      89
4     137

     ...
763   101
764   122
765   121
766   126
767    93
Name: Glucose, Length: 768, dtype: int64
<class 'pandas.core.series.Series'>
```
We can choose the columns of a dataframe by using square bracket notation. We have to input the column name in square brackets, and then all of the rows of those columns will be selected. One important point to note here is that a single column of a dataframe is a series object. Let us select two columns and see their object type:

Code:
```
print(df[['Glucose','Insulin']])
```

TABLE 4.4

Output of df.head() After Changing Column Names

	Pregnancies	Glucose	BP	ST	Insulin	BMI	DPF	Age	Outcome
0	6	148	72	35	0	33.6	0.627	50	1
1	1	85	66	29	0	26.6	0.351	31	0
2	8	183	64	0	0	23.3	0.672	32	1
3	1	89	66	23	94	28.1	0.167	21	0
4	0	137	40	35	168	43.1	2.288	33	1

```
print(type(df[['Glucose','Insulin']]))
```
Output:
```
     Glucose Insulin
0      148      0
1       85      0
2      183      0
3       89     94
4      137    168
..     ...    ...
763    101    180
764    122      0
765    121    112
766    126      0
767     93      0

[768 rows x 2 columns]
<class 'pandas.core.frame.DataFrame'>
```
We can pick two or more columns by passing a list of column names. In this example, two columns are of a dataframe object. Therefore, selecting more than one column produces a dataframe object containing the selected columns. For selecting rows, there is a method called ".iloc[]" which takes integers like the list's bracket notation.

Code:
```
#Selecting Rows
print(df.iloc[0]) #intiger index argumnets
print(type(df.iloc[0]))
print('\n-------\n')
print(df.iloc[0:2])#selecting two rows
print(type(df.iloc[0:2]))
print('\n-------\n')
print(df.iloc[0:2]['Glucose'])#selecting values in rows
```
Output:
```
Pregnancies     6.000
Glucose       148.000
BP             72.000
ST             35.000
Insulin         0.000
BMI            33.600
DPF             0.627
Age            50.000
Outcome         1.000
Name: 0, dtype: float64
<class    'pandas.core.series.Series'>
-------

Pregnancies Glucose BP ST Insulin  BMI  DPF Age Outcome
0         6    148 72 35       0 33.6 0.627  50           1
1         1     85 66 29       0 26.6 0.351  31       0
<class 'pandas.core.frame.DataFrame'>
-------

0     148
1      85
Name: Glucose, dtype: int64
```

We passed the index number for rows in the "iloc" function to choose them. As we previously observed, one-dimensional data is a series object in Pandas, and anything more than one-dimensional data is a dataframe object. More than one row can be selected by passing two indexes separated by a colon in the "iloc" function. The latter index will be exclusive, as it follows standard Python notation. Values of a particular row and column can be selected using the "iloc", followed by the column name, in bracket notation. This will work because "iloc" returns a series or a dataframe object, and we are already well-versed in selecting a column or list of columns from a dataframe.

Conditional Filtering in Dataframe

Supposing we have to filter out rows based on certain conditions - such as extracting rows where the BP of a patient is more than 110. We can apply such conditions to a column dataframe. Let us execute it below:

Code:

```
df[df['BP']>110]
```

Output:

In the code above, we are creating a dataframe where the values of the "BP" column are higher than 110 (Table 4.5). Therefore, it produced two columns matching our criteria:

```
df['BP']>
```

110' returns a Pandas series containing Booleans that satisfy our condition.)))

Code:

```
df['BP']>110
```

Output:

```
0      False
1      False
2      False
3      False
4      False
       ...
763    False
764    False
765    False
766    False
767    False
```

Name: BP, Length: 768, dtype: bool

The above output is a Pandas series of Booleans matching the criteria as we have noticed in NumPy arrays. We can also convert our dataframe into NumPy arrays. The example is shown in the next set of codes.

Code:

```
print(df.values)
print(type(df.values))
print(np.shape(df.values))
```

Output:

```
[[ 6. 148. 72…. 0.627 50. 1.]
```

TABLE 4.5

Filtering Rows based on Conditions on Columns

	Pregnancies	Glucose	BP	ST	Insulin	BMI	DPF	Age	Outcome
106	1	96	122	0	0	22.4	0.207	27	0
691	13	158	114	0	0	42.3	0.257	44	1

```
[ 1.  85.  66…. 0.351 31.  0.]
[ 8. 183.  64…. 0.672 32.  1.]…
[ 5. 121.  72…. 0.245 30.  0.]
[ 1. 126.  60…. 0.349 47.  1.]
[ 1.  93.  70…. 0.315 23.  0.]]
<class 'numpy.ndarray'>
(768, 9)
```

".values" returns a NumPy array with two dimensions (i.e. of 768*9 in size), as we know that the dataset has 768 rows and nine columns. Therefore, now we can perform all of the NumPy operations on this NumPy matrix.

Writing CSV Files from Pandas Dataframe

At this point, we have seen some operations on a dataframe. After accomplishing some data processing, we may be interested to save this dataframe for future use. We can save the dataframe into various formats like CSV, TSV, and Excel, among others - the same formats from which we can load a dataframe. In the succeeding example, we will study an example of writing a dataframe into the CSV format:

Code:

```
new_df = df[['Glucose','Insulin','BP']]
new_df.to_csv('small_dataset.csv')
with open('small_dataset.csv') as f:
    data = f.read()
print(data[:52])
print('-------------')
# exclution of index column
new_df.to_csv('small_dataset.csv',index=False)
with open('small_dataset.csv') as f:
    data = f.read()
print(data[:54])
```

Output:

```
,Glucose,Insulin,BP
0,148,0,72
1,85,0,66
2,183,0,64
-------------
Glucose,Insulin,BP
148,0,72
85,0,66
183,0,64
89,94,66
```

In the code above, we created a dataframe by selecting a few columns from our diabetes dataset. We selected the "Glucose", "Insulin", and "BP" columns and then used the ".to_csv()" instance method by creating a file name as an argument to the "new_df" dataframe. Next, we read the saved file and print a few characters to see the file format. In the first part, we observe an extra comma at the starting of the header row - which is a result of the index column - and is not a column name. To get rid of this index column, we can input "index=False" to the ".to_csv()" method as an additional argument. Therefore, in the next part, we can view that the index column is omitted.

Apply() Function

Dataframe has a method called ".apply()" which considers a function as an argument and applies the function to a dataframe or series. Supposing that we want to perform an operation on any column, and that operation is not available by default in the Pandas library, then we write a function for that particular operation and use the function in the ".apply()" method to that selected column. In the example below, we will categorize the age column into young, middle-aged, and aged:

Code:

```
def age_to_category(age):
    if age<18:
        return 'young'
    if age>=18 and age<=55:
        return 'middle-age'
    if age>55:
        return 'aged'
age = df['Age']
age.apply(age_to_category)
```

Output:

```
0     middle-age
1     middle-age
2     middle-age
3     middle-age
4     middle-age
        ...
763        aged
764   middle-age
765   middle-age
766   middle-age
767   middle-age
Name: Age, Length: 768, dtype: object
```

We have defined or written a function named "age_to_category" which takes a number, in this case - age, and returns the text where the condition satisfied. Readers may change the criteria according to their requirements in the Jupyter Notebook. The Jupyter Notebook provided with each of the chapters of this book is for practice. Once we defined the function, we selected the "Age" column, and, in the next line, we applied our custom function to that column. In the output, we notice that there are 767 values that have mentioned age. Now, let us add these values to our main dataframe and into a new column named "Age_cat".

Code:

Code:

```
df['Age_cat']=age.apply(age_to_category)
df.head()
```

Output:

We can create a new column by passing the name of the column in square brackets, similar to selecting a column and assigning values for each row. In the example above (Table 4.6), we are assigning values to the age categories and to the new column "Age_cat". By utilizing the ".head()"

TABLE 4.6

Output of df.head() to Verify the age_to_category Function

	Pregnancies	Glucose	BP	ST	Insulin	BMI	DPF	Age	Outcome	Age_cat
0	6	148	72	35	0	33.6	0.627	50	1	Middle-age
1	1	85	66	29	0	26.6	0.351	31	0	Middle-age
2	8	183	64	0	0	23.3	0.672	32	1	Middle-age
3	1	89	66	23	94	28.1	0.167	21	0	Middle-age
4	0	137	40	35	168	43.1	2.288	33	1	Middle-age

method, we can verify it as shown in the output. A new column is appended at the end after the "Outcome" column with age categories values. Therefore, the ".apply()" method has great purpose in dataframes for processing data and gives Pandas an edge over other spreadsheet programs.

Concatenating and Merging

Concatenating and merging tables are one of the fundamental operations for any spreadsheet program. In this section, we will study how to perform those tasks using Pandas methods. The following example is the concatenating two dataframes on a column basis or placing columns next to each other.

Code:

```
df1 = df[['Glucose','BP','ST']]
print(df1.head())
df2 = df[['Pregnancies','Insulin','BMI']]
print(df2.head())
concat_df=pd.concat([df1,df2],axis=1)
concat_df.head()
```

Output:

```
Glucose  BP ST
0    148 72 35
1     85 66 29
2    183 64  0
3     89 66 23
4    137 40 35
Pregnancies Insulin BMI
0           6       0 33.6
1           1       0 26.6
2           8       0 23.3
3           1      94 28.1
4           0     168 43.1
```

In the example above, we initially created two dataframes by selecting sets of three different columns from the main diabetes dataframe. The print ".head()" method is used for checking the structure of those dataframes. Then we used the "pd.concat()" method, which takes a list of dataframes to be concatenated as an argument. This method also considers arguments like "sort=False" and "axis=1". This sort of argument is used to arrange the columns alphabetically according to their name. The "axis" argument is 1 if we want to concatenate along the column (Table 4.7). While, if we want to add rows, then we can use "axis=0". This will append data along the row.

The next problem is interesting. Supposing we have two different tables, but we have a column and row between them - which is known as a key. We can merge these two tables on the basis of the common column value as shown in Figure 4.2:

TABLE 4.7

Concatenating two dataframes

	Glucose	BP	ST	Pregnancies	Insulin	BMI
0	148	72	35	6	0	33.6
1	85	66	29	1	0	26.6
2	183	64	0	8	0	23.3
3	89	66	23	1	94	28.1
4	137	40	35	0	168	43.1

Roll No.	Student Name
001	Suresh
002	Rohan
003	Niraj
004	Ravi

Roll No.	Marks
004	80
002	40
001	85
003	70

Merged

Roll No.	Student Name	Marks
001	Suresh	85
002	Rohan	40
003	Niraj	70
004	Ravi	80

FIGURE 4.2 Merger of Two Different Tables Based on a Common Key Value.

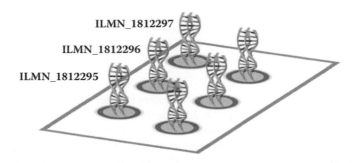

ILMN_1812297
ILMN_1812296
ILMN_1812295

FIGURE 4.3 Typical Microarray with Location IDs.

Figure 4.2 contains two tables on the top. The left one contains roll numbers and the names of students, and the other table indicates roll numbers and the marks obtained by students with these roll numbers. Therefore, we can merge these tables on the basis of roll numbers - which is the common field in both the tables, subsequently assigning the marks corresponding with the respective student names.

In the following example, we will map the microarray data with annotation data.

Figure 4.3 shows a typical gene expression microarray setup. Each location has a unique tag corresponding to a specific gene. When mRNAs are allowed to hybridize with this microarray, each location gives out a signal indicative of over- or under-expression of these mRNAs under specific conditions. The data, thus generated, is passed through the various steps of data processing, analysis, and comparison between control and treatment results in differentially expressed genes. We label this file as the expression file. The expression files contain three columns - the location ID, log fold change value, and the p-value. We also have an annotation file where we have location IDs, their respective gene symbols, and the titles or names. Let us map the gene expression data to gene symbols and names based on location identifiers.

Code:

```
expression = pd.read_csv('expression.csv')
```

TABLE 4.8

Gene expression data merging with gene symbols and names

	ID	Gene.symbol	Gene.title	logFC	P.Value
0	ILMN_1812297	CYP26B1	cytochrome P450 family 26 subfamily B member 1	2.887204	0.000011
1	ILMN_2392803	COL11A1	collagen type XI alpha 1 chain	2.587258	0.000047
2	ILMN_1730612	DBNDD2	dysbindin domain containing 2	-1.892913	0.000077
3	ILMN_1780349	TPR	translocated promoter region, nuclear basket p...	2.618084	0.000082
4	ILMN_1677636	COMP	cartilage oligomeric matrix protein	2.890289	0.000088

```
print(expression.head())
annotation = pd.read_csv('annotation.csv')
print(annotation.head())
merged_table = pd.merge(annotation,expression,on = 'ID')
merged_table.head()
```
Output:
```
                ID logFC P.Value
0 ILMN_1812297 2.887204 0.000011
1 ILMN_2392803 2.587258 0.000047
2 ILMN_1730612 -1.892913 0.000077
3 ILMN_1780349 2.618084 0.000082
4 ILMN_1677636 2.890289 0.000088
ID Gene.symbol                                     Gene.title
0 ILMN_1812297     CYP26B1 cytochrome P450 family 26 subfamily B member 1
1 ILMN_2392803      COL11A1 collagen type XI alpha 1 chain
2 ILMN_1730612      DBNDD2 dysbindin domain containing 2
3 ILMN_1780349          TPR translocated promoter region, nuclear basket
4 ILMN_1677636        COMP cartilage oligomeric matrix protein
```
We have provided the two table files in CSV format and labeled them as "expression.csv" and "annotation.csv", which are the expression file and annotation file, respectively. These files are first loaded as dataframes, and we used the ".head()" method to show the structure of the files. "pd.merge()" regards two dataframes as arguments and a parameter "on" which takes the column name which is common in both the file, then it merges the two files by mapping the data according to the key (i.e. common) column of both files. Lastly, the ".head()" method is used to produce the resultant dataframe (Table 4.8). We can see that the IDs are mapped with gene symbols, names, and expression values.

The goal of this chapter was to discuss the essential concepts of NumPy and Pandas, so that readers can get accustomed to these terminologies and their applications. The next chapter contains a number of colorful images, as it is about data visualization.

Exercise

1. Why is NumPy preferred over Python's nested list for multidimensional arrays?
2. Use NumPy to generate two arrays with the shapes (100 × 6) and (6 × 1) of random numbers sampled from a standard normal distribution.
3. Perform matrix operations like transposition and multiplication with the generated arrays.
4. Apply simple statistic operations on the array of shape (100 × 6).

5. Import "pandas" and open the "heart.csv" files supplied in the supplementary file.
6. Check the head of the dataframe above and grab the column names.
7. Select the columns with the name and find the mean using the ".mean()" function for every column.
8. Use "iloc[]" on the dataframe to select all of the columns, except for the last one.
9. In the second column (i.e. the column named "sex") indicates "1" for males and "0" for females. Use the apply method to change the map with "1" as "male" and "0" as "female".

5

Python for Data Visualization

Introduction

"Data science" is a buzzword in today's age of high throughput biology. When we say data science, we handle enormous amounts of data and arrive at insights into biological findings. Up until this point, we have learned how to handle large datasets and how to do an efficient calculation on these. Data visualization is another way to derive insights from data through visualizations by using elements like graphs (e.g. scatterplots, histograms, etc), maps, or charts that allow for the understanding of complexities within the data by identifying local trends or patterns, forming clusters, locating outliers, and more. Data visualization is the preliminary step after loading the data to view the distribution of values. Cleaning the data, checking the quality of data, doing exploratory data analysis, and presenting data and results are some of the necessary tasks that a data scientist needs to do before applying any Machine Learning or statistical model on the data. In this chapter, we will describe one of the primary data visualization libraries of Python called "Matplotlib" and draw a few basic graphics. Next, we will browse through a library called "Seaborn" which provides a high-level interface for drawing beautiful and informative statistical graphs. Lastly, we will learn about interactive and geographical data plotting.

Matplotlib

Matplotlib is the most popular plotting library in the Python community. It gives us control over almost every aspect of a figure or plot. Its design is familiar with Matlab, which is another programming language with its own graphical plotting capabilities. The primary goal of this section is to go through the basics of plotting using Matplotlib. If we have Anaconda distribution, then we have acquired Matplotlib installed by default, or else we have to install it using a "pip" installer. Matplotlib is imported as "plt", similar to "np" for NumPy and "pd" for pandas. To view the graphs in the Jupyter Notebook, we have to use a Jupyter function "%matplotlib inline" in the notebook. The notebook of this chapter can be found with the supplementary files labeled as "Python for Data Visualization5.ipnyb".

First, we need to possess data in order to visualize it. In this example, we have data of cell growth rate of two cell lines, as illustrated by "bg01" and "wa07". The data consists of a fold change of cell counts observed in six days. This is an example of data, and we will create it as a list of numbers, as shown below:

Code:
```
bg01 = [0.00,26.70,69.89,176.14,448.30,590.91]
wa07 = [0.00,21.88,126.56,438.28,706.25,840.63]
days = [1,2,3,4,5,6]
```

There are two ways to create Matplotlib plots - functional methods and object-oriented methods. The functional method is rather straightforward and simple, so we will begin learning that first.

Matplotlib Functional Method

Code:

```
plt.plot(days, bg01, 'k') # 'k' is for black color
plt.xlabel('Days post ploting')
plt.ylabel('Fold Change')
plt.title('BG01 Growth curve')
plt.show()
```

Output:

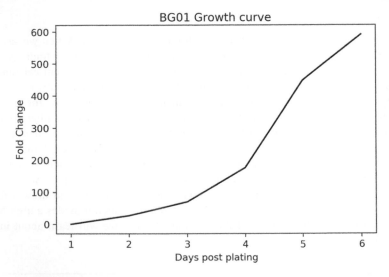

FIGURE 5.1   Simple Matplotlib Output.

The "plt.plot()" function takes various arguments, like "X" and "Y" values, that we can produce lists as well as np arrays for X and Y variables. In this example, we illustrated "days" in the X-axis and "bg01" in the Y-axis. Another variable we can provide is for color, and here we passed "k" which represents black (Figure 5.1). We will discuss the coloring, designing, and styles in detail later in this section. Next, we used axis label functions to name the axes and the plt.title() function for adding a title to the plot. Lastly, the plt.show() will show us the final plot. We have observed that we can use the different functions of Matplotlib to add various characteristics to our plot. Lastly, we have to apply the ".show()" function to determine the results. That is the main crux of the Matplotlib function type approach.

According to our requirement, we can draw two plots in the same canvas using a function called ".subplot".

Code:

```
# plt.subplot(nrows, ncols, plot_number)
plt.subplot(1,2,1)
plt.plot(days, bg01, 'k--')
plt.ylabel('Fold Change')
plt.title('BG01 Growth curve')# More on color and line options letter
plt.subplot(1,2,2)
plt.plot(days, wa07, 'k*-')
plt.ylabel('Fold Change')
plt.title('WA07 Growth curve')
plt.tight_layout()
plt.show()
```

Output:

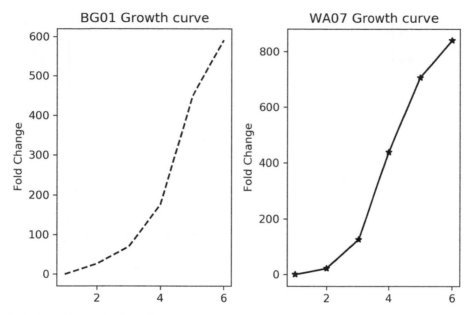

FIGURE 5.2   Two Plots on the Same Canvas.

The ".subplot" takes three arguments - the row number, the column number, and the plot number. We can think of this as dividing our canvas per the number of rows and columns. If we prefer to divide the canvas horizontally into two parts, then we have one row and two columns (Figure 5.2). Therefore, inputting ".subplot(1,2,1)" instructs it to carry out operations on the canvas with one row and two columns - selecting plot number 1 or the first from the left side. Similarly arguments (1,2,2) means selecting the second plot from a single row and double co-lumned canvas. After selecting the plot, we can use the functions to plot our data - such as labeling, for example - and then select other plots. In the example above, we customized the style of lines to show the differences (e.g. -- and *-). Moreover, we can also change the colors.

Matplotlib Object-Oriented Method

We have studied a functional method to plot figures. Now we will learn about the object-oriented method. The object-oriented method gives more flexibility than the functional method, and things improve more. Also, it is the most preferred and advisable way of using Matplotlib.

The idea of the object-oriented method is that first, we will create a blank figure object and name it "fig", and then we will add features like axes, plot data, labels, titles, and others to that figure object.

Code:

```
# Create Figure (blank space)
fig = plt.figure()
# Add set of axes to figure
axes = fig.add_axes([0.1, 0.1, 0.8, 0.8])
# left, bottom, width, height (range 0 to 1)
plt.show()
```

Output:

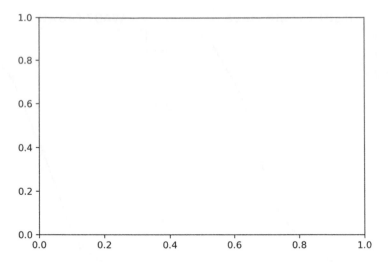

FIGURE 5.3   Creating a Blank Figure with Axes Using a Matplotlib Object-Oriented Method.

The first statement, "fig=plt.figure()" creates a blank space or an imaginary canvas - which is a figure object. In the second statement, the "fig.add_axes()" method considers four arguments, which are between 0 and 1, and define the percentage/ratio of space required for plotting or the location where we want to draw our axes. In this example (Figure 5.3), we have passed [0.1,0.1,0.8, and 0.8] for left, bottom, width, and height, respectively. Readers can play around with these numbers in the provided notebook to get a more clear picture of the functionalities. Next, we will add attributes to this axis object such as the plot, legends, and the title of the plot.

Code:

```
# Create Figure (blank space)
fig = plt.figure()
axes = fig.add_axes([0.1, 0.1, 0.8, 0.8])
# Add set of axes to figure
axes.plot(days, bg01, 'k')
axes.set_xlabel('Days post plating')
axes.set_ylabel('Fold Change')
axes.set_title('BG01 Growth curve')
plt.show()
```

Output:

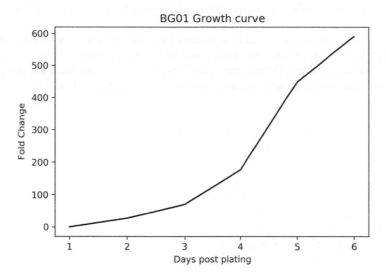

FIGURE 5.4   Adding a Plot and Other Attributes to the Axes Object.

Subsequently after creating a blank figure with axes, the ".plot()" method will draw the plot with a given list of *x*- and *y*-coordinates. The labels and the title were added using the ".set_xlabel", ".set_ylabel", and the "set_title" methods. The above result (Figure 5.4) is the same as Figure 5.1, where we used the function approach. Let us study how to use .subplots() in an object-oriented method.

Code:
```
fig, axes = plt.subplots(nrows = 1, ncols=2)
print(axes)
plt.show()
```

Output:

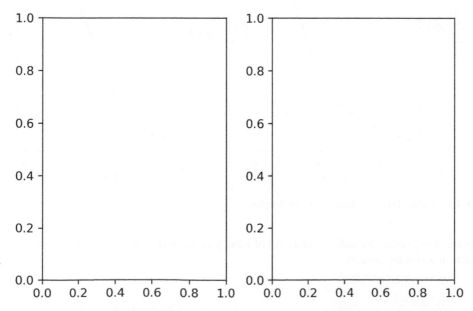

FIGURE 5.5   Calling "plt.subplots()" Will Return a Figure Object and Axes Objects.

```
[<matplotlib.axes._subplots.AxesSubplot object at 0x000001D58ECD66A0>
 <matplotlib.axes._subplots.AxesSubplot object at 0x000001D58ED30FD0>]
```

We can observe that "plt.subplots()" takes arguments for segmenting the canvas into numbers of rows and columns (Figure 5.5). It returns two objects, one of which is a figure object, and the other is the number of plots we have created. Printing the "axes" object shows us that there are two plots in it, and we can grab plots via indexes to add data and attributes to these (Figure 5.6).

Code:
```
fig, axes = plt.subplots(nrows = 1, ncols=2)
axes[0].plot(days, bg01, 'k--')
axes[0].set_xlabel('Days')
axes[0].set_ylabel('Cell Count')
axes[0].set_title('Growth Curve of BG01')
axes[1].plot(days, wa07, 'k*-')
axes[1].set_xlabel('Days')
axes[1].set_ylabel('Cell Count')
axes[1].set_title('Growth Curve of WA07')
# Display the figure object
fig.tight_layout()
plt.show()
```

Output:

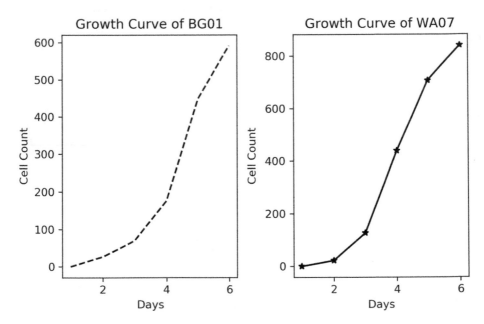

FIGURE 5.6 Adding Data and Attributes to the Subplots.

Plots are chosen using the indexes, and a list of *x* and *y* values is drawn on plots. Also, labels and titles are added in a similar fashion.

Resolution and Saving Figures

Through specifying the figure size, aspect ratio, and DPI, Matplotlib allows more flexibility for building custom plots using the "figsize" and "dpi" arguments. The "figsize" is a tuple expressed in inches of the figure's width and height, while dpi is the dots-per-inch - which is also the unit of resolution. Since we did not mention the "figsize" and the "dpi" in the previous cases, Matplotlib used its default values.

Code:
```
fig = plt.figure(figsize = (2,3), dpi=100)
axes = fig.add_axes([0.1, 0.1, 0.8, 0.8])
plt.show()
fig = plt.figure(figsize = (3,2), dpi=100)
axes = fig.add_axes([0.1, 0.1, 0.8, 0.8])
plt.show()
```

Output:

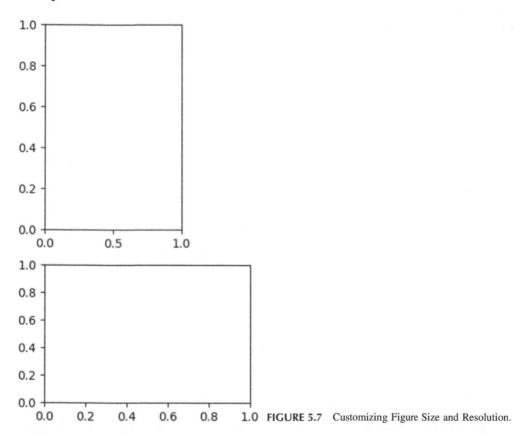

FIGURE 5.7 Customizing Figure Size and Resolution.

We have two figures of different sizes. The same can be done while initiating the subplots object by passing the same arguments - that is, the figsize and dpi to ".subplots()". Therefore, we can produce high-quality plots in terms of size and resolution (Figure 5.7). Matplotlib has functions to save these plots in several formats like PNG, JPG, SVG, PDF, and so on using the "fig.savefig()" method.

Code:
```
#Saving figures
fig.savefig("filename.png")
fig.savefig("filename.png",dpi = 200)
```

Appending this method after creating the plot will save the figure in the desired format. We have to indicate the name of the file along with the format as in "filename.png". The file will be saved in the ".png" format in the current working directory. We can also use the "dpi" argument here (dpi = 200), for saving high-resolution figures.

Legend

The graph legend describes the details of the data in the *Y*-axis of the graph. A graphical legend usually appears in a box on either side of the graph. Legends are required when we plot two or more types of data in a single frame to compare or illustrate different results. The box includes small pieces of every color and a brief description of the meaning of each color so that an observer should understand the meaning of colors and shapes in the chart in regard to our data. The graphic legends have always been an essential component of any graph report. Let us practice how to add a legend to a plot and customize its location too (Figure 5.8).

Code:
```
fig = plt.figure()
# Add set of axes to figure
axes = fig.add_axes([0.1, 0.1, 0.8, 0.8])
# Plot on that set of axes
axes.plot(days, bg01, label = 'BG01')
axes.plot(days, wa07, label = 'WA07')
axes.set_xlabel('Days post ploting')
axes.set_ylabel('Fold Change')
axes.set_title('Growth curve')
axes.legend(loc = 0)
plt.show()
```

Output:

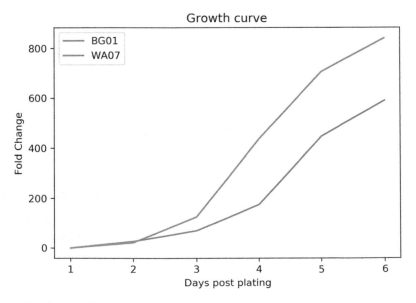

FIGURE 5.8   Adding the Legend to a Graph.

For setting legends, we have to use the "label" parameter while plotting our data or operate the "axes.plot()". The label is the description of the data which we want to show in the legend. Next, adding the statement "axes.legend()" will position the legend in the plot. This method works "loc" as an argument, which used to control the placement of the legend in the plot. This can also take location strings as arguments, as shown in Table 5.1.

TABLE 5.1

Strings and Respective Integers for the "loc" Argument of the "axes. legend()" Method

Location String	Location Code
'best'	0
'upper right'	1
'upper left'	2
'lower left'	3
'lower right'	4
'right'	5
'center left'	6
'center right'	7
'lower center'	8
'upper center'	9
'center'	10

Strings have their native sense, except for the "best" or code "0", which helps the legend method to position the legend box in an optimal place without or with only minimal overlap within the plot. In this example, we can notice that the graphs are colored, which is the default case in Matplotlib, and we did not select the black (i.e. "*k*") color. In the next section, we will further discuss this in detail.

Customization of the Plot Appearance

Malpotlib gives us a lot of flexibility while customizing the appearance of the plot. We can customize lines in terms of colors, line style, line width, transparency, and more. We will begin with line colors (Figure 5.9).

Code:
```
import numpy as np
fig = plt.figure()
ax = fig.add_axes([0.1, 0.1, 0.8, 0.8])
ax.set_xlabel('X axis') # Notice the use of set_ to begin methods
ax.set_ylabel('Y axis')
a = np.array([0,1,2,3,4,5,6])
ax.plot(a, a**2, color = "black")
ax.plot(a, a*2, color = "#F60707")                # RGB hex code
ax.plot(a, a + 3, color = "#0723F6", alpha=0.5) # half-transparant
```

Output:

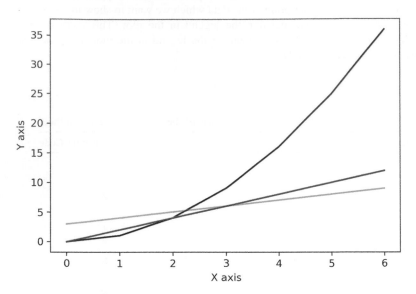

FIGURE 5.9   Customizing Line Colors.

The ".plot()" methods take an argument called color. We can use strings of a single character, color name, or color hex codes. We can use the following strings and characters for defining colors in Matplotlib (Table 5.2).

TABLE 5.2

Abbreviation of Available Colors in Matplotlib

Character	Color
"b"	Blue
"g"	Green
"r"	Red
"c"	Cyan
"m"	Magenta
"y"	Yellow
"k"	Black
"w"	White

We can also use an RGB hex code such as "#F60707", which allows us more flexibility to choose almost any color. The plot method takes another argument called "alpha" which denotes the transparency of the line and can be between 0 and 1, where 0 is 100% transparent and 1% indicates zero transparency.

Matplotlib has Various Options for Line Styling – Solid Lines, Dashed Lines, Etc. (Figure 5.10).

Code:

```
## Line and marker styles
x=np.array([1,2,3,4,5,6])
fig, ax = plt.subplots(figsize=(8,4))
```

```
ax.set_xlabel('X axis')
ax.set_ylabel('Y axis')
ax.plot(x, x+1, color="red", linewidth=0.25)
ax.plot(x, x+2, color="red", linewidth=2.00)
# possible linestype options '-', '--', '-.', ':', 'steps'
ax.plot(x, x+5, color="green", lw=3, linestyle='-')
ax.plot(x, x+6, color="green", lw=3, ls=':')
# possible marker symbols: marker = '+'
ax.plot(x, x+12, color="blue", lw=3, ls='-', marker='+')
# marker size and color
ax.plot(x, x+15, color="blue", lw=1, ls='-', marker='o', markersize=8,
markerfacecolor='yellow')
plt.show()
```

Output:

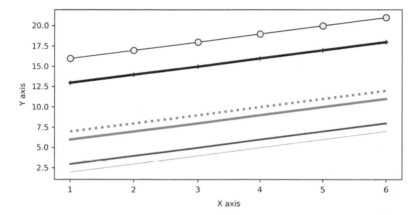

FIGURE 5.10   Customization of Plots for Line and Marker Styles.

Given below are some of the parameters accepted by the plot() method for customization of line styling, line width, and marker styling (Table 5.3). Markers are the coordinates of the given data - for example, the *x* and *y* points on the graph. The example code above shows the use of the mentioned parameters for plotting graphs.

TABLE 5.3

Line Styling Parameters

Style Parameter	Input Type
linestyle or ls	{'-', '--', '-.', ':', '', (offset, on-off-seq),…}
linewidth or lw	Float
Marker	marker style
markeredgecolor or mec	Color
markeredgewidth or mew	Float
markerfacecolor or mfc	Color
markerfacecoloralt or mfcalt	Color
markersize or ms	Float

Now that we have acquired enough information for building and customizing line plots, we will presently focus on some other types of specialized plots - such as scatterplots, histograms, box plots, and others. Most of these plot types are used in the Seaborn visualization library, which is a more advisable way for statistical plotting.

Scatterplot

A scatterplot is the simplest type of graph where data points are marked in the chart according to their vertical and horizontal coordinates. The "plt.scatter()" method produces a scatterplot, which uses x and y variables as parameters, along with optional marker styles (Figure 5.11).

Code:
```
plt.scatter(days,bg01)
plt.xlabel('Days')
plt.ylabel('Cell Count')
plt.title('Growth Curve of BG01')
```

Output:

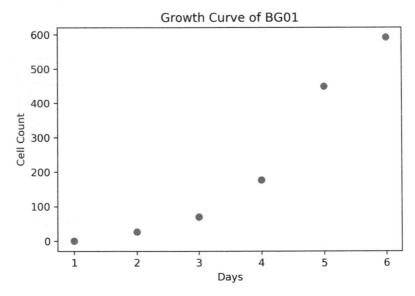

FIGURE 5.11   Scatterplot Using Matplotlib.

Histogram

Histograms provide the number of data points within a range of values for the visual interpretation of numeric data. Such a range of values is referred to as groups or bins. The height of the bar of a particular bin is its size. The histogram has large statistical applications in visualizing the spread and information of data.

For visualizing data, let us load our diabetes dataset from the previous chapter (Table 5.4).

Code:
```
import pandas as pd
df = pd.read_csv('datasets.csv')
df.head()
```

Output:

TABLE 5.4

The Diabetes Dataset

	Pregnancies	Glucose	Blood Pressure	Skin Thickness	Insulin	BMI	Diabetes Pedigree Function	Age	Outcome
0	6	148	72	35	0	33.6	0.627	50	1
1	1	85	66	29	0	26.6	0.351	31	0
2	8	183	64	0	0	23.3	0.672	32	1
3	1	89	66	23	94	28.1	0.167	21	0
4	0	137	40	35	168	43.1	2.288	33	1

Now, we will visualize the distribution of blood glucose among the observations using the "plt.hist()" method.

Code:

```
glucose = df['Glucose']
plt.xlabel('Blood Glucose Level')
plt.ylabel('Observations')
plt.title('Distribution of Blood Glucose Level')
plt.hist(glucose)
```

Output:

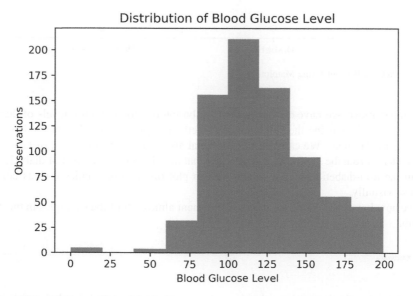

FIGURE 5.12 Plotting Histogram Using Matplotlib.

The histogram shows that the bin containing 100 as the blood glucose level is the most common, according to the data (Figure 5.12).

Boxplot

Boxplots contain five crucial descriptions of the data - specifically, the lowest value, lower quartile, median, upper quartile, and the highest value. These can be used to discover unusual observations like the outliers as well as to compare two or more distributions. The boxplot is sometimes also called a whisker plot.

Code:
```
diabetic = df[df['Outcome']==1]['Glucose']
non_diabetic = df[df['Outcome']==0]['Glucose']
plt.boxplot([diabetic,non_diabetic],labels = ['Diabetic','Non-Diabetic'])
plt.ylabel('Blood Glucose Level')
plt.show()
```

Output:

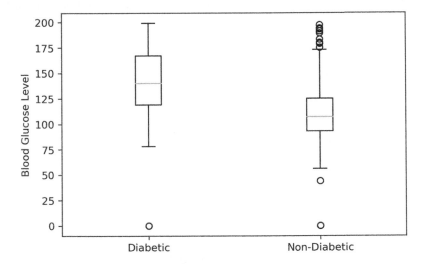

FIGURE 5.13   Plotting Boxplot Using Matplotlib.

In the example above, we have compared the distribution of blood glucose levels of diabetic versus non-diabetic observations using the "plt.boxplot()" method (Figure 5.13). This method considers a list of observations as the input. We can also pass optional arguments - such as a list of labels - and other styling parameters. From the graph, we can interpret that the blood glucose level of diabetic observation is more than the non-diabetic one. This way, we can plot distributions under various conditions and compare these visually.

Finally, we now have enough information to implement almost all of the examples in the gallery page of Matplotlib.

Seaborn

Seaborn is a data visualization package that is used to create attractive and detailed statistical plots with fewer lines of code as compared to Matplotlib. Seaborn has beautiful default styles and is also designed to work well with Pandas and NumPy libraries. Like Matplotlib, the Seaborn website also hosts a gallery page where various examples of statistical plots can be seen along with their implementation. In this section, we will cover a few statistical plots and the working interface of Seaborn. Seaborn can be installed using a "pip" or "conda" installer. Seaborn is imported as "sns", and the magic function "%matplotlib inline" will be required to view the plots in the Jupyter Notebook, since the package is based on Matplotlib.

Before plotting, we need to load data. Seaborn provides some example datasets. Among these, we will use the "iris" and "penguin" datasets. The "iris" dataset consists of three groups of 50 cases each, each of which refers to an iris plant type. For each plant observation, features such as sepal length, sepal width, petal length, and petal width are provided. The data for 344 penguins is included in the dataset "penguin". There are three different species of penguins collected from three islands in Antarctica, the Palmer, Archipelago. Observations of each case include penguin length, culmen depth, pinball length, body mass, and sex. All of the datasets are, by default, loaded as a Pandas dataframe. Let us begin with loading the "iris" dataset first.

Code:

```
import seaborn as sns
import pandas as pd
iris = sns.load_dataset('iris')
iris.head()
```

Output:

	sepal_length	sepal_width	petal_length	petal_width	species
0	5.1	3.5	1.4	0.2	setosa
1	4.9	3.0	1.4	0.2	setosa
2	4.7	3.2	1.3	0.2	setosa
3	4.6	3.1	1.5	0.2	setosa
4	5.0	3.6	1.4	0.2	setosa

Here is the appearance of the iris dataset.

Distribution Plots

Mostly, two types of plots are used to visualize the distribution of data. The first one is a histogram that we have already discussed, and the second one is kernel density estimation (kde). The "sns.distplot()" method of the Seaborn package allows us to plot these distribution plots. The "sns.distplot()" used many parameters, out of which pass data is compulsory, and other essential parameters include "kde" and "hist". They accept bool type data, which are used to plot either one of these or both of these. If we see the documentation of the "distplot()" method, then we can find parameters to customize our graph.

Code:

```
#histogram and kbe
fig, axes = plt.subplots(nrows=1, ncols=2,figsize=(8,4))
axes_0= sns.distplot(iris['sepal_length'],kde = True,
                     hist=True,color='black',bins = 8,ax=axes[0])
axes_0.set_title('Sepal length distribution with 8 bins ')
axes_1 = sns.distplot(iris['sepal_length'],kde = True,
                      hist=True,color='black',bins=9,ax=axes[1])
axes_1.set_title('Sepal length distribution with 9 bins')
plt.tight_layout()
plt.show()
```

Output:

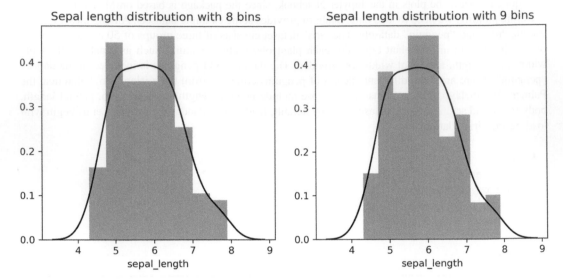

FIGURE 5.14   Distribution Plots Using Seaborn.

   In the example above, we have tried to visualize the distribution of sepal length among all types of iris plants. The kernel density estimation map shows the distribution of variables over a continuous interval, whereas the histogram illustrates the distribution at discrete intervals called bins. The shift in bins will, therefore, have an effect on the table, while kde does not depend on the bins and, rather, remains the same for the distribution of data - as shown in Figure 5.14. Therefore, kde is a better representation of the data distribution. Density plot peaks aid in displaying values distributed across the range.

Joint Plots

A scatterplot is best utilized in visualizing the correlation between two variables. On the other hand, Seaborn's joint plot method allows us to draw a scatterplot along with the distribution of each variable in their respective axes (Figure 5.15).

Code:
```
sns.jointplot(x ='petal_length', y ='petal_width',
            data = iris,kind='scatter',color='black')
```

Output:

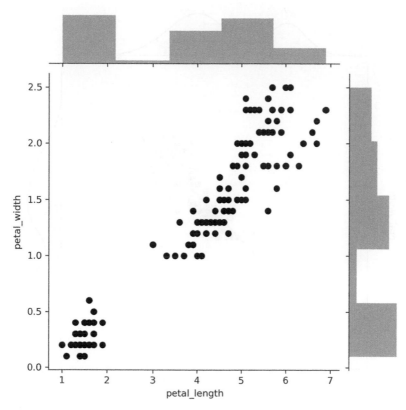

FIGURE 5.15 Joint Plot of Scatterplot and Axes Histograms.

The "jointplot()" method draws a combination of multivariable and univariable plots in the same figure, where a multivariable plot is the scatterplot and univariable plot is the distribution of the variables drawn on the axes (i.e. axes histograms). The arguments "*x*" and "*y*" are the variables, in this case, these are the column names "petal_length" and "petal_width" in the iris dataset, so the parameter "data" will consequently be "iris". The "kind" parameter takes arguments like "scatter", "reg", "resid", "kde", and "hex" for the respective type of graph. In the figure above, we can observe that the variable petal length and petal width are directly correlated. Now, we are going to input the "reg'" in the "type" parameter, which will give us a regression plot that resembles a scatterplot, except that Seaborn would actually draw a regression line on this. As far as the topic of Machine Learning is concerned, we have yet to actually discuss linear regression. However, later, as we approach the topic in the Machine Learning part of this book, we will explore how this line is actually built. Nevertheless and essentially, this only displays the scattered point data almost like a linear match, and we can actually see their linear relationship.

Code:
```
sns.jointplot(x ='petal_length', y ='petal_width',
        data = iris,kind='reg',color='black')
```

Output:

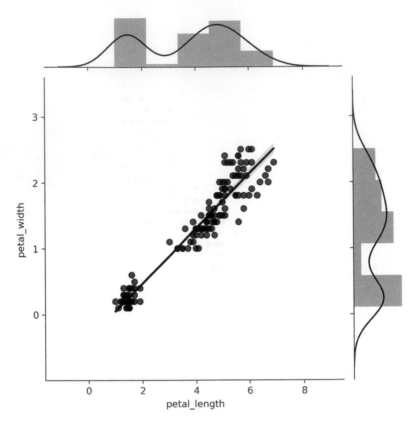

FIGURE 5.16   Regression Plot.

In Figure 5.16, we can see the estimated line which best fits the data. Furthermore, the graph also provides a density estimation of the distribution for variables, also known as the KDE plot. The slope of the regression line is positive; therefore, they can be considered as positively correlated.

Pairplot

The jointplot allows us to observe the relation between two variables. Similar to the jointplot, we have a method called the pairplot in Seaborn. The pairplot is essentially going to plot pairwise relationships for the numerical columns across an entire dataframe. It basically draws a jointplot among all of the numerical columns, which makes our work a lot more convenient rather than using the jointplot multiple times. We will now apply the penguins dataset, which we have discussed earlier in this chapter. Let us load the dataset and begin coding (Table 5.5).

Code:
```
import pandas as pd
penguins = sns.load_dataset('penguins')
penguins.head()
```

Output:

TABLE 5.5

Example Dataset – Penguins

	Species	Island	culmen_length_mm	culmen_depth_mm	flipper_length_mm	body_mass_g	sex
0	Adelie	Torgersen	39.1	18.7	181.0	3750.0	MALE
1	Adelie	Torgersen	39.5	17.4	186.0	3800.0	FEMALE
2	Adelie	Torgersen	40.3	18.0	195.0	3250.0	FEMALE
3	Adelie	Torgersen	NaN	NaN	NaN	NaN	NaN
4	Adelie	Torgersen	36.7	19.3	193.0	3450.0	FEMALE

As previously discussed, the penguins dataframe contains seven columns, as shown above. The head dataframe shows columns having "NaN" values, which means that this has no values and, hence, is not required for plotting graphs. Therefore, we will remove these "NaN" values by using the "dropna()" method of the Pandas dataframe. This will eliminate the rows having "NaN" values. After applying this method, our data is now ready for analysis.

The "sns.pairplot()" method uses a dataframe as a compulsory argument. In this case, it is penguins.

Code:
```
sns.pairplot(penguins)
```

Output:

With only a one-line statement, we gained these correlation plots among all of the numeral columns. Plotting this will take some time, depending on the number of numerical columns in our dataframe. In Figure 5.17, we can observe that almost all of the variables show some correlations. Among these, "flipper length" and "body mass" seem to be the most positively correlated. We can add more information to the graph by assigning categorical variables like "species" or "sex", which can be achieved by using a parameter called "hue". "Hue" will color the data points according to their category. Let us execute this.

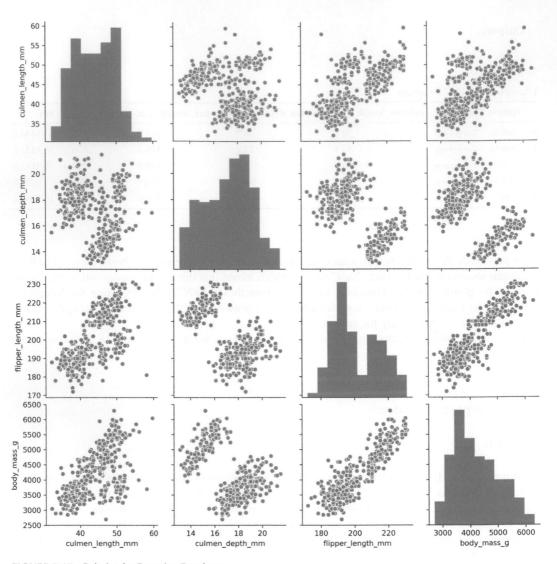

FIGURE 5.17   Pairplot for Penguins Dataframe.

Code:
```
sns.pairplot(iris,hue = 'species')
```

Output:

The potential "hue" parameter divides the data points by category or class, which provides more details than the correlations - including the distribution of variables and the correlation between the variables of their class, thereby providing deeper insights into our dataset. In the event that we must build an algorithm to classify penguins according to their species, then we can decide and prioritize which features are more important for the task. Figure 5.18 shows that "blue" and "green" classes (i.e. the species Adelie and

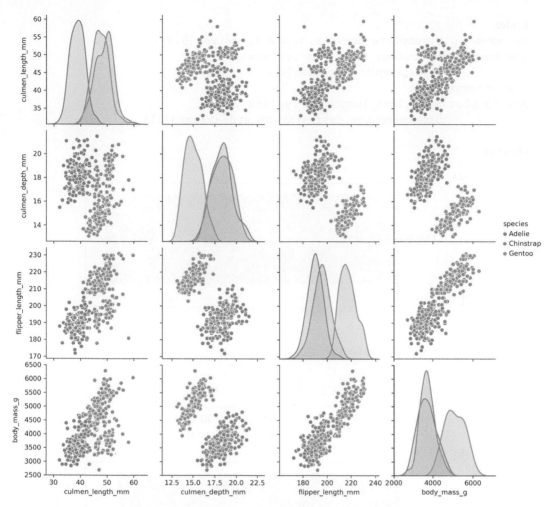

FIGURE 5.18 Pairplot with "Hue" Parameter.

Gentoo) are almost completely dissimilar or have non-overlapping data points. On the other hand, the "orange" class (i.e. the Chinstrap) is difficult to distinguish as it overlaps with both species in terms of their respective characteristics or variables. Moreover, we can distinguish all three species based on the scatterplot of culmen length with other variables. This allows us to do some feature engineering, where a form of mixed variables can be generated in order to obtain better results from Machine Learning models, which we will further discuss in the Machine Learning section of this book.

Barplot

A barplot is used to compare categorical or discreet variables or features. Categorical variables, as we learned in the previous chapter, are discrete values that usually have a fixed number of values and do not have any internal ordering. These are also known as nominal variables. In barplots, rectangle bars show the central tendency such as the mean, median, mode, and more of a variable for their respective category or categories.

Code:

```
fig, axes = plt.subplots(nrows=2, ncols=1,figsize=(5,8))
sns.barplot(x='culmen_length_mm',y='sex',
            data=penguins,ax=axes[0])
import numpy as np
sns.barplot(x='culmen_length_mm',y='sex',
            data=penguins,estimator=np.std,ax=axes[1])
```

Output:

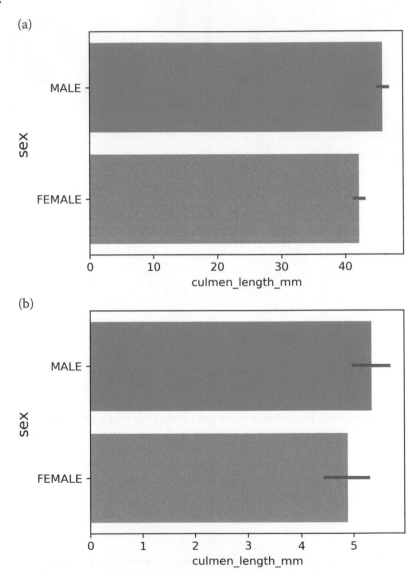

FIGURE 5.19 Barplot Examples: (a) Showing Mean as a Central Tendency, Which Is by Default, and (b) Has Standard Deviation as an Estimator.

In the above example, we divided the canvas by plotting two graphs. The method "sns.barplot()" takes *x* and *y* variables. The categorical variable here is sex, which states male and female as its two categories and is represented in the *y*-axes. The *x*-axes describe culmen length, and the central tendency or estimator (i.e. the basis for comparison) is the mean, which is set as default in the barplot method.

Figure 5.19a indicates that the length of male culmen is, on average, longer than the length of the female culmen. We may change the estimator measurement by using a function or a method as a parameter in the "estimator", as shown as an example in the second plot where the NumPy standard deviation method is passed as an estimator. Figure 5.19b illustrates that the lengths of the male's culmen have more standard deviation than the female's culmen. We can set the *y*-axis for the species in order to view the differences in culmen lengths among the species.

Boxplot

Boxplots and their uses have already been discussed in the Matplotlib section. In this part, we discuss Seaborn's boxplot method. The example is given below:

Code:

```
fig, axes = plt.subplots(nrows=1, ncols=2,figsize=(10,4))
sns.boxplot(x='culmen_length_mm',y='sex',
            data=penguins,ax=axes[0])
sns.boxplot(x='culmen_length_mm',y='sex',
            data=penguins, hue = 'species', palette="coolwarm",ax=axes[1])
fig.tight_layout()
```

Output:

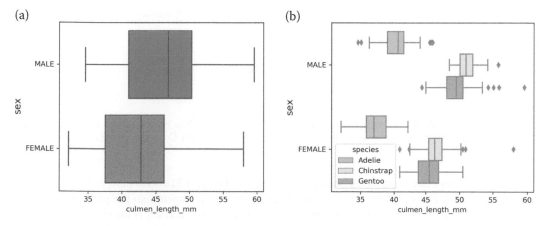

FIGURE 5.20 Seaborn Boxplot Method: (a) Has a Simple Boxplot and (b) Has a More informative Boxplot with Hue and Coloring.

Figure 5.20a shows a boxplot with default styling parameters. In this example, *x* is the culmen length and the category of the *y-axes* is sex, and this gives us the same insights as with the barplot, with additional information like quantiles and the spread of the variables. In Figure 5.20b, we have identified a specified species in the hue parameter. Moreover, the graphs, therefore, show more information on the pattern of culmen length variable with respect to sex and species. The Adelie species has a relatively small culmen length in comparison to the other two species. We have also introduced a styling parameter called the palette where we indicated "coolwarm" as the input, which drastically changed the default color theme of Seaborn. Colors are very important elements when plotting figures, as these can uncover the hidden patterns in data if used strategically. Seaborn has various contrasting color themes called color palettes so that we do not have to be preoccupied with the coloring aspects of our figure. "Coolwarm" is one of the most-used color palettes. Other interesting palettes like "colorblind", where the colors are colorblind-friendly, among others, can be found in its documentation. We can also set palettes in the Seaborn figure by using the "sns.set_palette()" method and through passing the required palette before the plotting statements.

Violin Plot

Violin plots are like boxplots, where instead of the mean, median, and quantiles, it shows the kernel density estimation or, specifically, the distribution of the variables on both sides of a line, which indicates the range of the variable.

Code:
```
fig, axes = plt.subplots(nrows=1, ncols=3,figsize=(12,4))
sns.violinplot(x='culmen_length_mm',y=' species',
               data=penguins,ax=axes[0])
sns.violinplot(x='culmen_length_mm',y='species',
               data=penguins, hue = 'sex',ax=axes[1])
sns.violinplot(x='culmen_length_mm',y='species',
               data=penguins, hue ='sex',split=True,ax=axes[2])
fig.tight_layout()
```

Output:

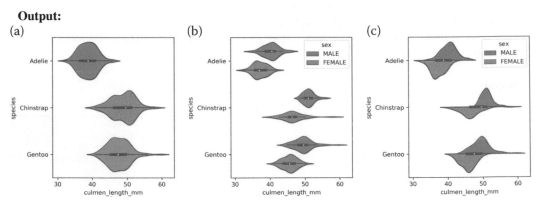

FIGURE 5.21   Violin Plots with Seaborn: (a) Simple Violin Plot, (b) Violin Plots Where Culmen Length Is Compared Based on Two Categorical Variables, Species and Sex, and (c) Secondary Category Comparison on Each Side of the Line.

Figure 5.21 shows three customizations of the violin plots. Figure 5.21a shows the violin plot with the default settings. In Figure 5.21b, "sex" is configured as the hue parameter, so the plots are torn apart according to the secondary category. Now, the species is the primary, and sex is the secondary category. As a result, the violin plot for the species is divided into six categories and are placed next to each other. An interesting aspect of the violin plot is that it has two halves which show exactly the same distributions and take up twice the amount of space. Therefore, instead of putting two violin plots next to each other, we could include one half showing the distribution of males and the other half showing the distribution of females of each species. We can do this by passing "split" equals "true" as an argument, and then in the process of splitting up the distributions, we are able to yield more information (Figure 5.21c). Now, we can directly compare the distributions based on two categories and also how the groups themselves on the x-axis compare with each other.

Heatmaps

Heatmaps are two-dimensional matrix visualization graphs that use color intensities. A heatmap is extremely useful for the graphical representation of the magnitude of an event in two dimensions. It helps to discover trends in real-time analytics and offers deep insights into the phenomena. Both axes of the heatmaps should have numerical values. Heatmaps locate their applications in visualizing correlations among various numerical variables after the calculation of the correlation matrix. Heatmaps are also used for visualizing changes in gene expression in expression data analysis similar to a microarray or RNA-seq analysis.

For the construction of the heatmap, we will calculate the correlation matrix. A correlation matrix is a matrix of the coefficient of correlations among the variables. The coefficient can be between −1 to 1, where −1 means negatively or inversely correlated, while 1 means positively correlated, and 0 means there is no correlation among the variables.

Code:
```
penguins.corr()
```

Output:

TABLE 5.6

Penguins Data's Correlation Matrix

	culmen_length_mm	**culmen_depth_mm**	**flipper_length_mm**	**body_mass_g**
culmen_length_mm	1.000000	−0.228626	0.653096	0.589451
culmen_depth_mm	−0.228626	1.000000	−0.577792	−0.472016
flipper_length_mm	0.653096	−0.577792	1.000000	0.872979
body_mass_g	0.589451	−0.472016	0.872979	1.000000

For a dataframe, there is just one statement to calculate the correlation matrix, and Table 5.6 shows the matrix of correlation coefficients between variables. Next, we will plot the heatmap for this correlation matrix (Figure 5.22).

Code:
```
sns.heatmap(penguins.corr(),annot=True,
            cmap='coolwarm',linewidth=1,linecolor='white')
```

Output:

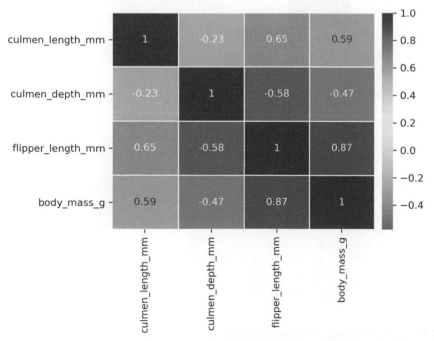

FIGURE 5.22   Heatmap of the Correlation Matrix from the Penguin Dataset.

Seaborn's "heatmap()" takes various parameters where the data matrix is compulsory, while the other parameters are for adding the annotations (i.e. the numerical values that will be shown inside the cells. Some styling parameters are present - like the cmap, linewidth, and linecolor, among others, where "cmap" is the color theme of Matplotlib similar to palettes in Seaborn. As Seaborn is based on Matplotlib, it, therefore, supports Matplotlib's color themes. The "linewidth" and "linecolor" are for styling the borderlines of the cells.

Cluster Maps

Usually, a clustered heat map is created on variables of similar scales. When variables have different scales, a standard normalization is initially needed for the data matrix to be scaled. The order of rows is determined by conducting a hierarchical study of rows. Similar rows tend to be placed adjacent to one another in the plot. The sequence of the columns is similarly determined.

Code:
```
sns.clustermap(penguins.corr(),annot=True,
               cmap='coolwarm',linewidth=1,linecolor='white')
```

Output:

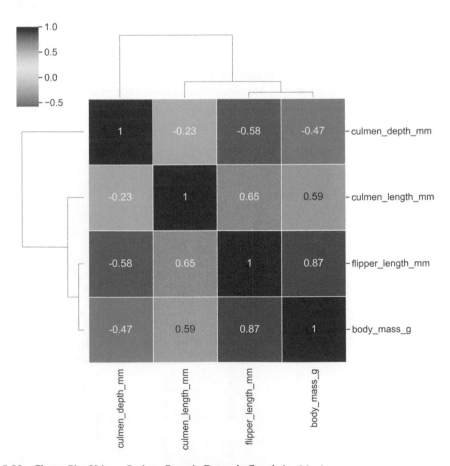

FIGURE 5.23   Cluster Plot Using a Seaborn Penguin Dataset's Correlation Matrix.

The "clustermap()" method takes the same parameters as the "heatmap()" method. The output is different, since the cluster map has a similarity grouping of variables, as shown in Figure 5.23 in the axes. The figure clearly shows that the body mass and the flipper length are the most similar, which corroborates with earlier results obtained in this section.

Regression Plot

Regression plots show the line that best fits the data and can be used to see the trend or even for the prediction of unknown values. Seaborn has a "regplot()" method for plotting regression plots, but here we will use the "lmplot()" method, which is a combination of the "regplot()" and FacetGrid class. The FacetGrid class allows us to view the distribution of a variable in several panels separately, along with its relationship with multiple variables within our dataset's subsets. The "lmplot ()" is more computationally intensive and is designed to fit regression models across a limited subset of data while providing a convenient interface. Next is the example of a simple regression plot using "lmplot()":

Code:
```
sns.lmplot(x='culmen_length_mm',y='body_mass_g',data=penguins)
```

Output:

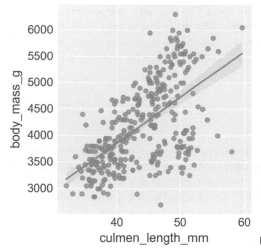

FIGURE 5.24 Seaborn "lmplot()" Method.

As observed in Figure 5.24, the best fit line or regression line is plotted along some margins, which are generally non-optimal lines plotted using the main data subsets. This gives us tolerable margins, along with the best fit line. The "lmplot()" can also take hue as a parameter (Figure 5.25).

Code:
```
sns.lmplot(x='culmen_length_mm',y='body_mass_g',
           data=penguins,hue='sex')
```

Output:

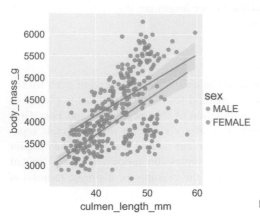

FIGURE 5.25   Seaborn's "lmplot()" with Hue as a Parameter.

Adding the hue parameter draws two regression plots in the same graph, plots for males and females. The plot shows that both lines are significantly apart from each other, and this gives us some insights about the relation between body mass and culmen length in regard to a different sex.

Seaborn is quite useful in creating easy and attractive data visualization. Parameters like palette or cmap allow for a variety of beautiful and contrasting color themes. Also, parameters, like hue, often provide more pieces of information in the graphs. The most commonly used method is pairplot, which gives a general image of the numerical variables. The best thing about Seaborn is that its methods are rather versatile, and almost every item can be configured in a single method or in a statement line.

Plotly – Interactive Data Visualization

Plotly is the library used for producing interactive graphs developed by a company known also by the same name - Plotly. Plotly allows the creation and hosting of interactive visualization in websites, and it also offers free offline hosting which we will use here. In this section, we will show the usage and features of Plotly in the Jupyter Notebook, along with Pandas' built-in visualization wrappers. Pandas visualization methods can be used for exploratory data analysis . In this section, we will show a simple example of plotting a histogram using Pandas, and then we will convert it into an interactive version. This section is to show the application of Plotly along with Pandas for creating interactive plots. Various Pandas visualization methods can be found in Pandas' visualization documentations. Some of the basic methods in Pandas include "bar" or "barh" for bar plots, "hist" for histogram, "box" for boxplot, "kde" or "density" for density plots, "scatter" for scatterplots, "pie" for pie plots, etc.

Code:
```
penguins['body_mass_g'].hist()
```

Output:

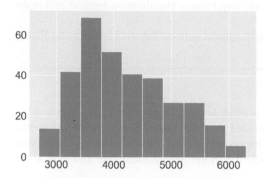

FIGURE 5.26 Plotting Histogram Using Pandas Visualization Methods.

Above is an example (Figure 5.26) of a histogram for body mass variable or column for the penguin dataframe. To make it interactive, we are required to install Plotly using the pip installer – "pip install plotly". Next, we have to set up the Pandas plotting backend as Plotly or instruct Pandas to use Plotly for plotting the graphs (Figure 5.27). We can do this by typing "pd.options.plotting.backend = "plotly"" in the Jupyter Notebook cell once (recall "%matplotlin inline"). Now, the plots will be interactive. Let us look at the examples.

Code:
```python
import pandas as pd
pd.options.plotting.backend = "plotly"
penguins['body_mass_g'].hist()
```

Output:

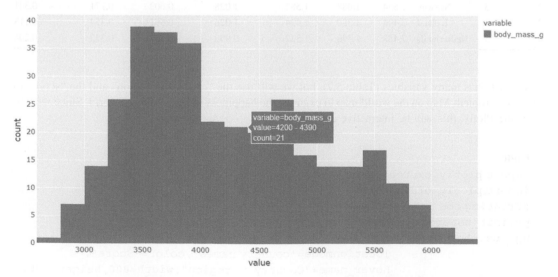

FIGURE 5.27 Interactive Histogram Using Plotly as Backend.

Interactive plotting entails such characteristics as we can hover over the image to see the respective values. Additionally, we have options for zooming, selecting an area, directly downloading or saving the image file, and much more. In the same way, we can use any Pandas visualization method and make these interactive.

In the next section, we will discuss the plotting of geographical data on the world map using Plotly.

Geographical Plotting

For plotting geographical data, we first need to acquire data that is spread across countries. A dataset file in "CSV" format named "happiness_2019.csv" is provided with the supplementary material of this chapter. This file contains variables associated with the happiness index of various countries for the year 2019, as surveyed by the United Nations. Let us load the file in the Pandas dataframe and look at the contents of the file.

Code:

```
happiness = pd.read_csv('happiness_2019.csv')
happiness.head()
```

Output:

TABLE 5.7

Happiness Index of Countries

	Overall Rank	Country or Region	Score	GDP per Capita	Social Support	Healthy Life Expectancy	Freedom to Make life Choices	Generosity	Perceptions of Corruption
0	1	Finland	7.769	1.340	1.587	0.986	0.596	0.153	0.393
1	2	Denmark	7.600	1.383	1.573	0.996	0.592	0.252	0.410
2	3	Norway	7.554	1.488	1.582	1.028	0.603	0.271	0.341
3	4	Iceland	7.494	1.380	1.624	1.026	0.591	0.354	0.118
4	5	Netherlands	7.488	1.396	1.522	0.999	0.557	0.322	0.298

The data has many variables (Table 5.7), but we will use the "country or region" and the "score" to plot a Choropleth Map on the world map to examine the happiness index across the world. Since we will be using Plotly, this will be interactive as well.

Code:

```
import plotly.express as px
import plotly.offline as py
#Creating the interactive map
py.init_notebook_mode(connected=True)
fig = px.choropleth(happiness, locations="Country or region",
                    locationmode='country names',color="Score",
                    hover_name="Country or region",width=800,height=800,
                    title='Happiness Index (2019) Across World')
#Showing the figure
py.offline.iplot(fig)
```

Output:

FIGURE 5.28   Interactive Geographical Plotting Using Plotly.

This process is remarkably easy with Plotly. We only have to pass the names of the columns that we want to visualize in a dataframe, and Figure 5.28 is subsequently created. Let us talk about the imports "plotly.express" and "plotly.offline". Plotly express is a wrapper over Plotly, just how Seaborn is over Matplotlib. Using Plotly as a backend for Pandas makes things more convenient to implement. Using Plotly from scratch is like using Matplotlib, where a few additional statements have to be written to get interactive plots. Therefore, Plotly Express has many methods, which makes our task as easy as using one-line statements, much like Seaborn. One such method is "choropleth" which is used to build interactive Choropleth Maps or geographical plotting. We have discussed that Plotly can be used for hosting plots or graphs. Moreover, it has two modes - online and offline - where offline is free to use. Therefore, the import "plotly.offline" is present to be able to plot offline. The statement "py.init_notebook_mode (connected = True)" allows the local connection of the Jupyter Notebook with the Plotly interactive platform, which basically plugs some javascript libraries in the notebook, and, as a result, we obtain an interactive output. The "choropleth" method takes a dataframe or dictionary object, and the "locations" argument creates the list of names of countries or states or places that should match with the "locationmode" values. Location mode can be customized by passing a GeoJSON file. GeoJSON files have coordinates, shapes, and the names of the regions. The "color" parameters take the list of variables according to the intensity of colors provided, which is the "Score" column of the happiness index table in our example. The "hover_name" provides information while hovering over the region on the map, which is a kind of interaction.

These all are basic requirements for building interactive geographical plots.

In this section, we have studied different libraries for data visualization as well as the types of graphs used to see patterns in our datasets. The graphs previously discussed are applicable to any kind of exploratory data analysis. For more chart types, we can go to the Seaborn or Matplotlib gallery pages, and, with the information in this chapter, we are ready to use any of the implementations. We have also seen a quick build-up of interactive visualization that is made interactive through the Jupyter Notebook by hovering over pieces of information, zooming, or the like.

Excellent and informative data visualization from start to finish is crucial for any data science project. Exploratory data visualization will provide us with more information about the data and may help users to select effective features and models. To summarize, data visualization will make it easy for us to communicate the results and make these understandable to the audience. In the next chapter, we will understand the concepts of the dimensionality of data, principal component analysis, and selecting features in Python.

Exercise

1. Create a NumPy array with "a = np.arange(0,50)" and plot "a**2" versus "a" using the Matplotlib functional method.
2. Use the objected-oriented method to plot the same array with "-" and a green-colored line.
3. Create two subplots in the same canvas and plot "a**2" versus "a" and "a**3" versus "a" in the respective subplots.
4. Save the above figure with 300dpi resolution.
5. Open the "seeds_dataset.csv" provided with the supplementary material with the Pandas dataframe. Use Seaborn's "joitntplot()" method to plot "lengthOfKernel" versus "widthOfKernel".
6. Draw a plot to see the correlation among the variables while using "seedType" as the hue.
7. Use a boxplot to view the distribution of "perimeter" among the three variables.
8. Find the correlation matrix of the dataframe and draw cluster maps on it.
9. Use the Pandas build in a scatterplot to view "lengthOfKernel" versus "widthOfKernel", while the Plotly packages are installed.
10. Plot the "GDP per capita" from the "happines_2019" dataset in geographical plotting and observe if this variable has any correlation with the happiness index.

6

Principal Component Analysis

Introduction

The number of features in a dataset is known as its dimensions; as the number of features increases, the number of dimensions increases. With the dawn of high-throughput technologies in the field of bio-technology - such as microarray, RNAseq, single-cell transcriptomics, etc. - large dimensional datasets are being generated. Often, high-dimensional data may pose several challenges, such as the following:

1. **The first challenge is an increase in the cost of computation.** It is apparent, that the more number of features there are, the greater the computational power is required for their analysis.
2. **The second challenge lies in the visualization of high-dimensional data.** We can visualize up to three-dimensional data, where two-dimensional data are most comfortable to plot and understand. In the previous chapter, we have primarily seen two-dimensional visualizations (i.e. the relationship between two variables, or sometimes three variables, using the hue parameter) which were for categorical values only.
3. **The final challenge is data redundancy.**

Such problems that arise due to high-dimensional data are called the curse of dimensionality, which is a very popularly used term in the data science community. There are two main approaches in dealing with this:

1. **The first is to drop or remove some features which are highly correlated.** However, the challenge of deciding the cut-off of correlation to exclude the features from the dataset and the corresponding loss of information due to their exclusion prevails.
2. **The second involves feature engineering.** This is the process of deriving new features from the available features. Therefore, if we have "n" number of features in the raw dataset, then we will create "d" number of features in our transformed dataset, where "d" $\ll$ "n".

One of the most widely used technique for dimensionality reduction without significantly losing information is called the principal component analysis (PCA).

Variance as Information

In statistics, the variance is a measure of how far a value is from the mean value in a dataset. It indicates how dispersed the values in a particular dataset are. Let us understand this further by using some examples. Supposing we count the number of hands of 100 individuals with the task of predicting the number of hands in the next individual encountered. If there are no congenital disabilities, then the plot of the data will be a straight line at 2, with no variance (Figure 6.1a). Now, let us proceed to another task that involves recording the everyday average temperature for 100 days, and then trying to predict the next day's temperature. An example of this data is shown in Figure 6.1b. Now, this kind of data is something we would like to explore and build models on and calculate mean, median, range, and other values, since it has variations. Therefore, while making predictions, the principle behind making predictions or building models is that the variance is preserved. With this much information, we are one step closer to PCA.

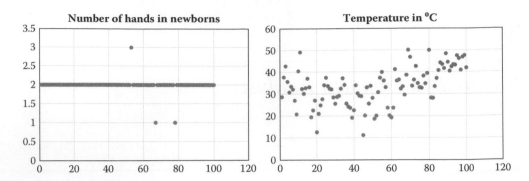

FIGURE 6.1   (a) Observations for the Number of Hands in 100 Individuals and (b) the Recorded Average Temperature for 100 Days.

PCA is a technique used for reducing the number of features - which is also known as dimensionality reduction - without significant loss of information or, specifically, by preserving the variance.

Data Transformation

In PCA, we construct "d" features from raw "n" features that allow us to preserve maximum variance, where "d" < "n". It may be noted that "d" and "n" are mutually exclusive or, precisely, the features "d" are new and are not a subset of "n" features.

Supposing we have a dataset that has four features or dimensions, as expressed below (Table 6.1):

After transforming these features, the dataset will be reduced to only two dimensions. (Table 6.2):

These two features are the linear combination of four raw features, such that:

$$P_1 = a_1f_1 + a_2f_2 + a_3f_3 + a_4f_4$$

$$P_2 = b_1f_1 + b_2f_2 + b_3f_3 + b_4f_4$$

In this example, a's and b's are eigenvectors, and these form the transformed matrix "**W**" - which, in turn, transforms the raw feature space into the new space where the variance is maximum (Table 6.3). In terms of a matrix, this can be explained as: (Table 6.4),

TABLE 6.1

Raw Features of a Dataset

f_1	f_2	f_3	f_4
5.1	3.5	1.4	0.2
4.7	3.2	1.3	0.2

TABLE 6.2

Transformed Features of the Dataset

P_1	P_2
2.64	−5.20
2.45	−4.77

TABLE 6.3

Transformed Matrix, **W** with Eigen Vectors

a_1	b_1
a_2	b_2
a_3	b_3
a_4	b_4

TABLE 6.4

Raw Features of the Data F

f_1	f_2	f_3	f_4

Therefore, $P = W * F$, or specifically, the new features, are dot products of eigenvectors and raw features.

Case Study

Figure 6.2a shows an example of relations between two variables with variations. As the goal of PCA is to preserve the maximum variations, we can, therefore, calculate the direction of variations - as shown in Figure 6.2b. The length of the directed lines corresponds to the magnitude of variation. In simple terms, the long arrow points towards more variance, and the short arrow points towards lesser variance. We know that variations are informative, so, in this example, the long arrow is a more informative component as compared to the shorter one. Next, we will transform the whole axis towards the directions of the variances, and the new coordinates will be calculated using the new axis. Hence, we will produce the new values of the features with preserved variance. These new coordinates are called transformed versions of the raw ones, and, in order to calculate these new coordinates, we require a transformation matrix, as is shown above. This transformation matrix is composed of eigenvectors and eigenvalues. Eigenvectors are the direction of the change (i.e. the axes), and eigenvalues are the magnitudes of change and are used for plotting coordinates. In Figure 6.2c, we can observe the maximum spread of data in the P_1 direction, along with a little variation in P_2, so here we can use only P_1 values which have the maximum information and is made up of F_1 and F_2 - thereby reducing the dimensions form two to one. This concept of dimensionality reduction may be applied for any number of dimensions.

Now, we will study PCA step-by-step using Python and its visualization. We will use the iris dataset as a case study here. Let us load the requisite libraries and the iris dataset (Table 6.5).

PCA Feature Transformations

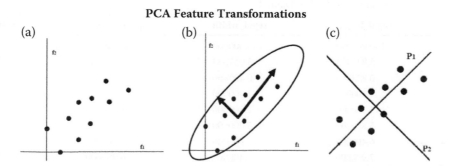

FIGURE 6.2 PCA Feature Transformation: (a) Scatterplot Between Two Features Having Variances, (b) Direction of Maximum Variations, and (c) Changed Axis Towards Maximum Variation.

Code:
```
import pandas as pd
import seaborn as sns
import numpy as np
import matplotlib.pyplot as plt
%matplotlib inline
iris = sns.load_dataset("iris")
iris.head()
```

Output:

The iris dataset, as discussed in chapter 5, is a dataset of three iris species with the sepal length, sepal width, petal length, and petal width as its features. Here, we have four features, so it is four-dimensional data. We will apply PCA to these four features and reduce the dimensions from four to two while retaining the maximum variance. Pandas has a method called ".describe()" which shows preliminary statistics for every feature using just one statement.

Code:
```
iris.describe()
```

Output:

Table 6.6 shows count or particularly, the number of instances, mean value, standard deviations, and quantile for each of the features in the iris dataset.

TABLE 6.5

Iris Dataset

	sepal_length	sepal_width	petal_length	petal_width	Species
0	5.1	3.5	1.4	0.2	setosa
1	4.9	3.0	1.4	0.2	setosa
2	4.7	3.2	1.3	0.2	setosa
3	4.6	3.1	1.5	0.2	setosa
4	5.0	3.6	1.4	0.2	setosa

TABLE 6.6

Description of Raw Features of the Iris Dataset

	sepal_length	sepal_width	petal_length	petal_width
Count	150.000000	150.000000	150.000000	150.000000
Mean	5.843333	3.057333	3.758000	1.199333
Std	0.828066	0.435866	1.765298	0.762238
Min	4.300000	2.000000	1.000000	0.100000
25%	5.100000	2.800000	1.600000	0.300000
50%	5.800000	3.000000	4.350000	1.300000
75%	6.400000	3.300000	5.100000	1.800000
Max	7.900000	4.400000	6.900000	2.500000

PCA: Step-by-Step

Standardization of the Features

In Table 6.6, we can observe that the data statistics values are very different for all four features - such as the means and the standard deviation. This shows that the data is of varying scale and that the features need to be normalized or brought to the same scale. The method used to do this is called standardization - where the data is transferred to a standard scale. The standard scale is where the mean is zero and the standard deviation is one for each feature. This can be achieved by subtracting the mean by each instance value of a feature and then dividing it by the standard deviation. The result is also called the z score, and the data is converted into a normal distribution.

Code:
```
X = iris.iloc[:,0:4]
Y = iris.iloc[:,4]
iris_std = (X-X.mean())/X.std()
iris_std = pd.concat([iris_std,Y],axis=1)
iris_std.head()
```

Output:
To standardize the features, we have first separated the features "X" from labels "Y". Second, we subtracted the means of the features with their respective instances and then divided this by the standard deviation. The standardized values finally obtained were concatenated with the labels to complete the dataset. The description of the standardized features of the iris dataset is obtained as follows (Table 6.7):

Code:
```
iris_std.describe()
```
Output:

In order to see that the mean is almost near zero, and the standard deviation is assigned as one for each of the features (Table 6.8).

Let us view the data spread and relations with and without standardization using visualization.

TABLE 6.7

Standardized Iris Dataset

	sepal_length	sepal_width	petal_length	petal_width	Species
0	−0.897674	1.015602	−1.335752	−1.311052	setosa
1	−1.139200	−0.131539	−1.335752	−1.311052	setosa
2	−1.380727	0.327318	−1.392399	−1.311052	setosa
3	−1.501490	0.097889	−1.279104	−1.311052	setosa
4	−1.018437	1.245030	−1.335752	−1.311052	setosa

TABLE 6.8

Description of Standardized Features of the Iris Dataset

	sepal_length	sepal_width	petal_length	petal_width
Count	1.500000e+02	1.500000e+02	1.500000e+02	1.500000e+02
Mean	−1.515825e−15	−1.823726e−15	−1.515825e−15	−8.526513e−16
Std	1.000000e+00	1.000000e+00	1.000000e+00	1.000000e+00
Min	−1.863780e+00	−2.425820e+00	−1.562342e+00	−1.442245e+00
25%	−8.976739e−01	−5.903951e−01	−1.222456e+00	−1.179859e+00
50%	−5.233076e−02	−1.315388e−01	3.353541e−01	1.320673e−01
75%	6.722490e−01	5.567457e−01	7.602115e−01	7.880307e−01
Max	2.483699e+00	3.080455e+00	1.779869e+00	1.706379e+0

Code:
```
fig, axes = plt.subplots(nrows=2, ncols=2,figsize=(12,12))
axes_0= sns.distplot(iris['sepal_length'],kde = True,
                     hist=True,color='black',ax=axes[0,0])
axes_0.set_title('Sepal length distribution raw data')
axes_1 = sns.distplot(iris_std['sepal_length'],kde = True,
                      hist=True,color='black',ax=axes[0,1])
axes_1.set_title('Sepal length distribution standardize data')
axes_2 = sns.scatterplot(x='sepal_length',y='sepal_width',
                         data = iris,hue='species',s=60,ax=axes[1,0])
axes_2.set_title('Sepal length vs Sepal width raw data')
axes_3 = sns.scatterplot(x='sepal_length',y='sepal_width',
                         data = iris_std,s=60,hue='species',ax=axes[1,1])
axes_3.set_title('Sepal length vs Sepal width standardize data')
plt.tight_layout()
plt.show()
```

Output:
In Figure 6.3, the plots on the left are those of raw data, and the ones on the right are of the standardized data. We can observe that the data has been centered at zero, but the distribution has been preserved. The same can be seen for the scatterplot: the data is centered at zero, but the variations have been preserved. Therefore, the information has been maintained, but the data is scaled, and now we can do data operations. Now, we will discover the direction of maximum variance.

Obtain the Eigenvectors and Eigenvalues

Since it is our goal to preserve the variance, then we will use the covariance matrix to produce the eigenvectors and eigenvalues - specifically, the direction and magnitude of the variances.

Code:
```
x_std = iris_std.iloc[:,:4].values # features
y = iris_std.iloc[:,4].values # labels
cor_mat = np.corrcoef(x_std.T)
eig_vals, eig_vecs = np.linalg.eig(cor_mat)
print('Correlation Matrix \n{}'.format(cor_mat))
print('\nEigenvectors \n{}'.format(eig_vecs))
```

```
print('\nEigenvalues \n{}'.format(eig_vals))
```

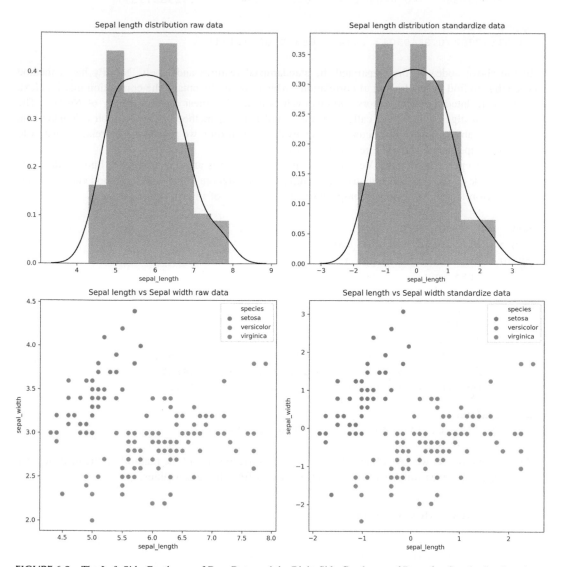

FIGURE 6.3 The Left-Side Graphs are of Raw Data, and the Right-Side Graphs are of Data after Standardization; Axes Are Made Darker to Show the Differences.

Output:

```
Correlation Matrix
[[ 1.             -0.11756978     0.87175378      0.81794113]
 [-0.11756978     1.             -0.4284401      -0.36612593]
 [ 0.87175378    -0.4284401       1.              0.96286543]
 [ 0.81794113    -0.36612593      0.96286543      1.         ]]

Eigenvectors
[[ 0.52106591    -0.37741762     -0.71956635      0.26128628]
 [-0.26934744    -0.92329566      0.24438178     -0.12350962]
```

```
[ 0.5804131      -0.02449161     0.14212637     -0.80144925]
[ 0.56485654     -0.06694199     0.63427274      0.52359713]]
```

```
Eigenvalues
[2.91849782 0.91403047 0.14675688 0.02071484]
```

In the above code, we first separated the standardized features and labels. NumPy has a method ".corcoef()" to find the coefficient of correlation - which essentially entails the correlation matrix. Next, we have calculated the eigenvectors and eigenvalues using the linear algebra library of NumPy. The NumPy linear algebra library is called "linalg'", and it has a method "egn()" which calculates and returns two matrices - the eigenvectors and eigenvalues. Therefore, we passed the correlation matrix to the method "np.linalg.eig()". The resultant values are printed.

We can also perform Singular Vector Decomposition, which is another method to find the direction and magnitude of the covariance matrix. The NumPy linear algebra library has a method called "svd()" used to perform singular vector decomposition. This is a type of matrix factorization.

Code:
```
u,s,v = np.linalg.svd(cor_mat)
print('Eigenvectors \n{}'.format(u))
print('\nEigenvalues \n{}'.format(s))
```

Output:
```
Eigenvectors
[[-0.52106591   -0.37741762     0.71956635     0.26128628]
[  0.26934744   -0.92329566    -0.24438178    -0.12350962]
[ -0.5804131    -0.02449161    -0.14212637    -0.80144925]
[ -0.56485654   -0.06694199    -0.63427274     0.52359713]]
```

```
Eigenvalues
[2.91849782 0.91403047 0.14675688 0.02071484]
```

As we can observe, we have obtained the same values irrespective of the method. In the next step, we have to select the axes which have maximum variances from the eigenvalues.

Choosing Axes with Maximum Variance

Code:
```
# list of eigenvalue and eigenvector pairs
eig_pairs = [(np.abs(eig_vals[i]), eig_vecs[:,i]) for i in range(len
(eig_vals))]

# Sort the (eigenvalue, eigenvector) tuples from high to low
eig_pairs.sort()
eig_pairs.reverse()

# printing the sorted list
print('\nEigenvalues and eignevector pairs in descending order:')
for i in eig_pairs:
print(i)
```

Output:
```
Eigenvalues and eignevector pairs in descending order:
```

```
(2.9184978165319975, array([ 0.52106591, -0.26934744, 0.5804131, 0.56485654]))
(0.9140304714680688,    array([-0.37741762,    -0.92329566,    -0.02449161,
-0.06694199]))
(0.14675687557131487,    array([-0.71956635,    0.24438178,    0.14212637,
0.63427274]))
(0.020714836428618374,    array([  0.26128628,   -0.12350962,   -0.80144925,
0.52359713]))
```

The magnitude of the eigenvalue determines the amount of the variance in the axis. Therefore, sorting the eigenvalues will provide us with the axis that has maximum variation. The code above does the same and prints the eigenvalues in descending order, with corresponding eigenvectors. This is done to confirm the order visually. We can also plot these values, according to the amount of variance, in each axis.

Code:
```
tot = sum(eig_vals)
var_exp = [(i / tot)*100 for i in sorted(eig_vals, reverse=True)]
cum_var_exp = np.cumsum(var_exp)
fig = plt.figure()
ax = fig.add_axes([0,0,1,1])
PC = ['PC1', 'PC2', 'PC3', 'PC4']
plot = ax.bar(PC,var_exp)
ax.plot(cum_var_exp, 'r*-')
def autolabel(plot):
"""Text label above each bar in *rects*, displaying its height."""
for rect in plot:
height = rect.get_height()
ax.annotate('{}'.format(round(height,2)),
xy=(rect.get_x() + rect.get_width() / 2, height),
xytext=(0, 3),  # 3 points vertical offset
textcoords="offset points",
ha='center', va='bottom')
plt.ylabel('Preserved Variation')
autolabel(plot)
plt.show()
```

Output:
Each eigenvalue and its respective vectors form a principal component. The variance, in terms of percentage, is calculated and plotted. In Figure 6.4, we can note that the first two principal components compromises more than 95% of the data. Based on variance data, we can select these two principal components as axes. This will preserve > 95% of variance from the four features of the iris dataset. Afterwards, we will create the transformation matrix - which will transform our instances to the co-ordinates of our new principle components.

Let us look at the construction of the transformation or the projection matrix using Python:

Code:
```
matrix_w = np.hstack((eig_pairs[0][1].reshape(4,1),
eig_pairs[1][1].reshape(4,1)))
print('Matrix W:\n', matrix_w)
```

Output:
```
Matrix W:
[[  0.52106591 -0.37741762]
```

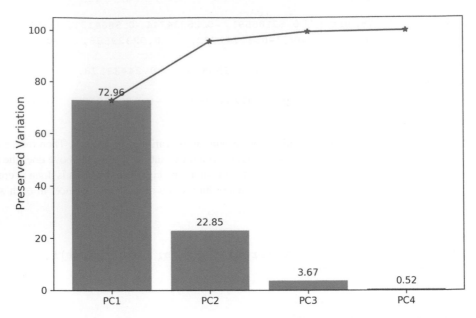

FIGURE 6.4   Plot of the Magnitude of Variance in Each Principal Component and Cumulative Variances.

```
[-0.26934744  -0.92329566]
 [ 0.5804131   -0.02449161]
 [ 0.56485654  -0.06694199]]
```

In the code above, we have selected the eigenvectors for the top two eigenvalues in terms of variation content. The matrix we acquired is the corresponding W matrix, which was introduced in Table 6.3 - the eigenvector matrix. This matrix will transform the coordinates of feature values or project them into the new coordinate system of principal components. The next and final step is to transform the original features in the dataset to principal components.

Here is the method to transform the original dataset or projection:

Code:
```
features= iris.iloc[:,0:4].values
iris_transform = features.dot(matrix_w)
iris_transform = pd.DataFrame(iris_transform,columns = ['PC1','PC2'])
iris_transform = pd.concat([iris_transform,Y],axis=1)
iris_transform.head()
```

Output:

	PC1	PC2	Species
0	2.640270	−5.204041	setosa
1	2.670730	−4.666910	setosa
2	2.454606	−4.773636	setosa
3	2.545517	−4.648463	setosa
4	2.561228	−5.258629	setosa

Transforming the features means multiplication or the dot product of features with the matrix **W** to obtain the transformation or projection matrix. In the code above, the first statement captures the scaled features of the iris dataset. In the next statement, the dot product is calculated using the NumPy method "dot()", and we obtain the resultant transformed matrix. This matrix is then concatenated with the labels, and, hence, we obtain reduced features for our iris dataset. Now, we can visualize the whole dataset in two dimensions.

Code:
```
sns.set_style("whitegrid")
sns.scatterplot(x = 'PC1',y = 'PC2',data = iris_transform,hue = 'species',s = 60)
```

Output:

Figure 6.5 displays a plot to visualize the whole iris dataset in two dimensions while preserving a 95.81% variance. Visually, we can easily classify these species using these two principal components. We can even categorize the species based on only the first principal component while preserving a 72.96% variance (Refer Figure 6.4). As illustrated in Figure 6.5, the instances <−1 of PC1 are Setosa, between −1 and 1 are mostly Versicolor, and >1 are mostly Virginica. Therefore and finally to conclude, PCA is also known as dimensionality reduction.

Programing Drive

We have seen that PCA contains so many steps, but those steps are required to understand the important and underlying processes. Python has a library called Scikit-Learn, which is the most popular library for Machine Learning, and we learn about it in chapter 8 of this book. This library contains a method "PCA", which does all of the steps internally and provides us with results in just one or two statements. This library comes with the Anaconda distribution, and it can also be installed using the "pip" or "conda" installer.

Code:
```
from sklearn.decomposition import PCA
sklearn_pca = PCA(n_components=2)
print(x_std.shape)
PCs = sklearn_pca.fit_transform(x_std)
print(PCs.shape)
```

Output:
```
(150, 4)
(150, 2)
```
Scikit-Learn in Python is termed as "sklearn". Scikit-Learn, or sklearn's decomposition library, contains the PCA class. PCA class is initialized with "n_components" which is the number of top principals required in results. Here, we initiate it using "n_components=", where we require the top two principal components having maximum variance. This will reduce our datasets to that amount of engineered features. PCA requires a scaled data, so we will pass the "x_std" data, which is standardized and has 150 × 4 shapes (i.e. 150 rows and 4 columns or features). The "fit_transform()" method will take the scaled data and transform it into its principal components. Hence, as a result, we obtain a matrix size of 150 × 2 or, specifically, 150 rows and 2 columns, with reduced features or principal components. Let us plot them below:

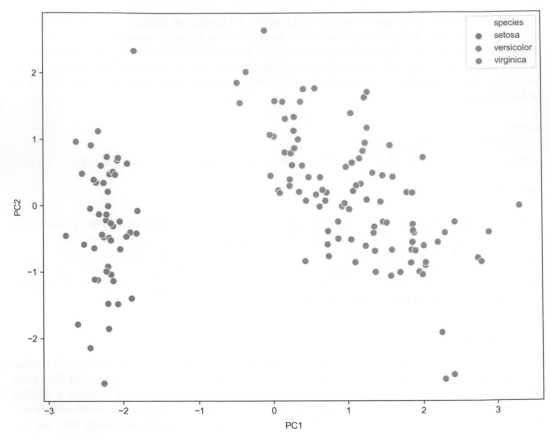

FIGURE 6.5 Plotting the Iris Dataset into Two-Dimensional Space Using the Principal Components, where Each Colored Dot is a Species (Setosa in Blue, Versicolor in Orange, and Virginica in Green).

Code:
```
iris_transform = pd.DataFrame(PCs,columns=['PC1','PC2'])
iris_transform = pd.concat([iris_transform,Y],axis=1)
fig, axes = plt.subplots(figsize=(10,8))
sns.set_style("whitegrid")
sns.scatterplot(x='PC1',y='PC2',data
= iris_transform,hue='species',s=60)
```

Output:

The plot in Figure 6.6 is similar to the plot in Figure 6.5, which was obtained by the stepwise calculation of PCA. Therefore, PCA can be achieved by using only two statements with sklearn. Furthermore, we can get the amount of variance preserved in the components using the attribute "explained_variance_ratio_".

Code:
```
sklearn_pca.explained_variance_ratio_
```

Output:
```
array([0.72962445, 0.22850762])
```

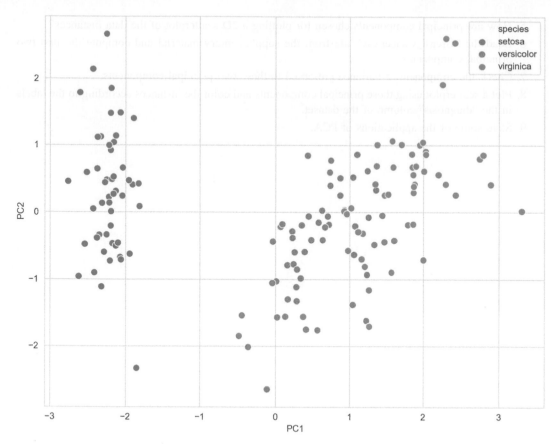

FIGURE 6.6 Plotting the Iris Dataset into Two-Dimensional Space Using the Principal Components Calculated Using Sklearn's PCA Class, where Each Colored Dot is a Species - where Blue is Setosa, Orange is Versicolor, and Green Is Virginica.

The output states that the first two principal components contain 72.96% and 22.85% variance, respectively.

PCA is an important tool for reducing dimensionality, which is an absolute necessity for data scientists. PCA makes Machine Learning algorithms faster, as fewer features take less time to compute. It aids with the visualization of high-dimensional data that we have seen in this chapter, and we will use this in the subsequent chapters as well. This helps to improve the performance of Machine Learning models by eliminating unwanted variances - which may otherwise also reduce the efficiency of the model. While modified versions of PCAs are available and applied, PCAs are also utilized in conjunction with other methods to obtain the desired results.

Exercise

1. Explain the importance of variation within data.
2. Why do we normalize data before performing PCA?
3. How is PCA associated with the variance of the dataset?
4. How are features transformed into principal components?

5. How are principal components chosen for plotting a 2D scatterplot of the data instances?

6. Load the "Breast_Cancer.csv" file from the supplementary material and compute the first two principal components.

7. Check the component of variance preserved in these two principal components.

8. Plot a scatterplot using these principal components and color the instances according to the labels in the "diagnosis" column of the dataset.

9. State some of the applications of PCA.

7

Hands-On Projects

In this chapter, we are going to do certain real-world data analyses using the knowledge we have gained so far. The first project involves the analysis of differential gene expression using microarray data, where we will discover up- and down-regulated genes, and then determine the biological process that may be influenced by these differentially expressed genes. The second project includes an analysis of the genotype-phenotype relationship, where we will use the SNP example data and predict the pharmacogenomics phenotypes.

Differential Gene Expression Analysis

The chain of nucleic acids is often considered as the language of life. The expression of this language designs an organism. Mostly, the stable form of genetic material is DNA, which encodes and stores most of the characteristics of the organism and passes this down from generation to generation. The central dogma of biology is fundamental to the existence of all living organisms. DNA is translated into RNA. mRNA is translated into proteins - which are the primary functional units for any cell, and these determine the fate of cells. Various environmental factors can have an effect on cells. All of these factors can be studied by comparing the expression of healthy cells and the affected cells - where healthy cells are called control cells and the affected cells are deemed treated cells. To measure gene expression, the microarray - a high-throughput technology for gene expression analysis - is used. Microarrays are gene chips where short single-stranded oligonucleotides which have a complementary sequence to genes are anchored on microscopic slides that are often referred to as gene chips or DNA chips. These oligonucleotides act as probes to detect gene expression. The RNAs are collected from the samples, converted into cDNAs, labeled with fluorescent dyes, and spread over the microarray chip. The main idea is the cDNA will be hybridized with the respective complementary probes by forming stable hydrogen bonds, and, hence, fluorescence can be detected at each spot. The intensity of the fluorescence is directly proportional to the number of RNAs in that spot, thus quantifying the expression of the respective gene (Figure 7.1).

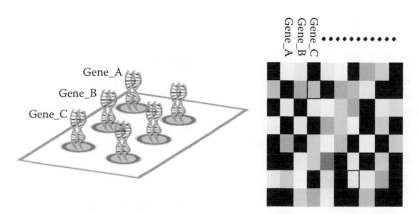

FIGURE 7.1   Microarray Chip Architecture and Resultant Intensities.

A high-resolution camera takes a picture of the chip, and different types of software are used to measure the intensities at the spots. Each chip contains more than 30,000 spots. Therefore, microarrays can be used to detect the expression of tens of thousands of genes instantaneously.

For differential gene expression analysis, we require a minimum of two sets of data - a control set and a treated set. The recorded intensities are then processed for background noise-filtering and are then recorded for further analysis. There are various bioinformatics resources, such as the Gene Expression Omnibus and ExpressArray, where the microarray data generated from multiple experiments are deposited and are publicly available. There are various pipelines to analyze microarray data, and any may be utilized based on research objectives. In this example, we will learn about the basic handling of microarray data and preliminary analysis. For this project, we have an example data set consisting of four control and four treated samples.

A typical microarray data analysis is comprised of the following steps:

- Feature extraction
- Quality control
- Normalization
- Differential expression analysis

The procedure discussed above - that is, of converting the image of microarray into significant intensities or quantifiable values that are saved in binary or text format - is known as feature extraction. Quality control (QC) of data is then carried out for ascertaining the quality of data. Because the intensity of emission from fluorescence tags also depends upon the intensity of the exposed light source, experimental artifacts may be produced due to the change of intensities of the exposed light. To minimize these artifacts, the distribution of the expressed genes in every sample is made equal. Let us study how this is done.

Quality Control

The first step in Python data analysis is always importing the required libraries.

Code:

```
import pandas as pd
import seaborn as sns
from sklearn.decomposition import PCA
import matplotlib.pyplot as plt
import numpy as np
%matplotlib inline
```

Using the libraries, we will do plottings, data handling, and PCA. In this part, we will load two datasets that will be available in the supplementary files with this chapter, along with the Jupyter Notebook. The supplementary file shows the intensities of the spots, and Supplementary File 2 is the annotation file where the information about the genes corresponding to the spots is provided.

Code:

```
expression = pd.read_csv('expression_data.csv')
print('Expression File Contains')
print(expression.head())
print(expression.info())
annotation = pd.read_csv('attotation_file.csv')
print('\n\nAnnotation File Contains')
print(annotation.head())
```

Output:

```
Expression File Contains
   ID_REF       Control_1    Control_2    Control_3    Control_4    Treated_1 \
0  1007_s_at       1752.5        986.7       1398.1       1107.1       1539.2
```

1	121_at	810.1	496.3	228.7	619.1	441.9
2	1316_at	958.5	918.9	531.2	1314.2	1113.7
3	1552257_a_at	517.1	691.2	375.6	563.8	216.4
4	1552264_a_at	1471.4	1408.7	1334.7	1923.2	1451.9

	Treated_2	Treated_3	Treated_4
0	1605.3	4663.0	1995.4
1	427.1	324.9	242.6
2	769.2	1182.8	974.8
3	260.1	446.2	291.4
4	1866.6	823.1	1367.5

```
<class 'pandas.core.frame.DataFrame'>
RangeIndex: 16525 entries, 0 to 16524
Data columns (total 9 columns):
```

#	Column	Non-Null Count	Dtype
---	------	--------------	-----
0	ID_REF	16525 non-null	object
1	Control_1	16525 non-null	float64
2	Control_2	16525 non-null	float64
3	Control_3	16525 non-null	float64
4	Control_4	16525 non-null	float64
5	Treated_1	16525 non-null	float64
6	Treated_2	16525 non-null	float64
7	Treated_3	16525 non-null	float64
8	Treated_4	16525 non-null	float64

```
dtypes: float64(8), object(1)
memory usage: 1.1+ MB
None
Annotation File Contains
```

	ID_REF	GB_ACC	Gene Symbol	ENTREZ_GENE_ID
0	1007_s_at	U48705	DDR1 /// MIR4640	780 /// 100616237
1	1053_at	M87338	RFC2	5982
2	117_at	X51757	HSPA6	3310
3	121_at	X69699	PAX8	7849
4	1255_g_at	L36861	GUCA1A	2978

As explained here, the column "ID_REF" refers to the spot ids, and the respective columns in the expression file represent the intensities of the samples for particular spots. The expression file contains the expression of 16,525 spots or genes. In the annotation file, the "ID_REF" column also exists, which is the spot IDs, and the respective columns are the identifiers of the corresponding genes.

The quality of data could be inferred by visualizing the distribution of data.

Code:

```
for i in expression.columns[1:]:
sns.distplot(expression[i], hist=False, label=i, axlabel ='Expression')
```

Output:

The distribution of the raw values does not clearly provide any information, as the range is very large, specifically from ~0–80,0000 (Figure 7.2). Comparing values in this range has many limitations, including the tendency of bias for calculations towards large numbers. One of the most common transformation method applied to microarray data is the logarithmic transformation.

Table 7.1 shows an example of raw values and $\log_2$ transformed values. In this table, we can see that the $\log_2$ transformation changes the range of 1–800,000 to around ~0–20. Now, we will learn how to perform a logarithmic transformation of the raw intensity data using Python.

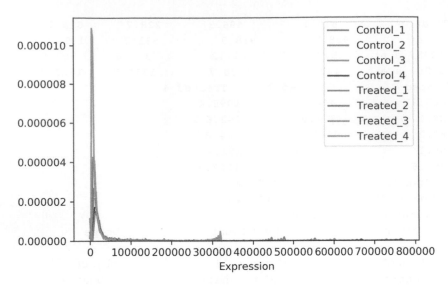

FIGURE 7.2 Distribution of Raw Values of the Intensities for Samples.

TABLE 7.1

Log2 Transformation of Numbers

Raw Values	Log2 Values
1	0
10	3.321928
100	6.643856
1000	9.965784
10000	13.28771
100000	16.60964
800000	19.60964

Code:
```
log_expression = np.log2(expression.iloc[:,1:])
for i in log_expression.columns:
    sns.distplot(log_expression[i], hist=False,label=i,
                         axlabel ='Log2 Expression')
```
Output:
Figure 7.3 illustrates the distribution of $\log_2$ transformed data, and here we can view the distribution of samples clearly. The distributions are almost comparable there, so the quality of data is satisfactory. Next, we will observe the bar pots to compare the quantiles of the sample data.

Code:
```
fig, axes = plt.subplots(figsize=(12,8))
ax = log_expression.boxplot(color='k')
                         ax.set_ylabel("Log2 Expression)
```
Output:
Boxplots show certain similar trends in the data with some deviations (Figure 7.4). We can observe that the means are different for the samples; thus, we have to normalize the data to overcome the experimental artifacts.

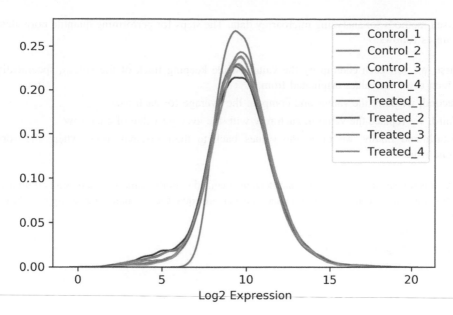

FIGURE 7.3 Distribution of Log2 Transform Data.

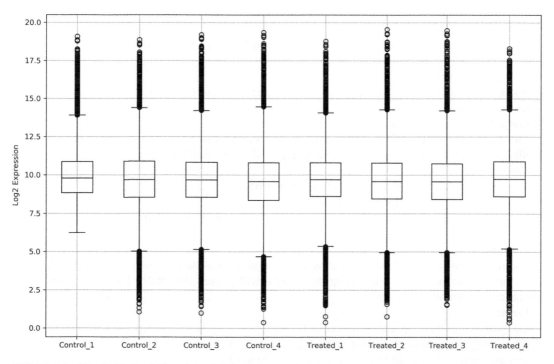

FIGURE 7.4 Boxplot for Log2 Transformed Values.

Normalization

To compare the data, we require the distributions to be as equal as possible without losing the variations among themselves. This process is called normalization. Quantile normalization is one of the most widely used methods for analyzing microarray data. The steps for performing quantile normalization in Python are as follows:

1. First, we sort each column by the values while keeping track of the values, specifically, from which rows the values originated from.
2. Second, we store the values and compute the average for each row.
3. Third, we replace the value in each row with the average value of each row.
4. Fourth and finally, we place the values back to their original rows, where they originally came from.

Therefore, this transforms the new data matrix as normalized with the same sum across all columns. Below, Python function uses a Pandas dataframe and can perform quantile normalization using the above steps.

Code:

```
def quantileNormalize(input):
    temp = input.copy()
    #compute rank
    dic = {}
    for col in df:
        dic.update({col : sorted(temp[col])})
    sorted_df = pd.DataFrame(dic)
    rank = sorted_df.mean(axis = 1).tolist()
    #sort
    for col in temp:
        t = np.searchsorted(np.sort(temp[col]), temp[col])
        temp[col] = [rank[i] for i in t]
    return temp
Norm_samples=quantileNormalize(expression.iloc[:,1:])
Norm_samples
```

Output:

The form table (Table 7.2) is rather difficult to understand in terms of the differences after normalization. Visualization of the distribution may provide some insights.

Code:

```
log_norm_expression = np.log2(Norm_samples)
for i in log_norm_expression.columns:
    sns.distplot(log_norm_expression[i], hist=False,label=i,
                axlabel = 'Log2 Normalizes Expression')
```

Output:

Figure 7.5 illustrates all of the distributions as identical and has stacked these one on top of the other after quantile normalization. Visualizing the boxplot will present the effect of normalization on the quantiles and central tendencies.

Code:

```
fig, axes = plt.subplots(figsize=(12,8))
ax = log_norm_expression.boxplot(color='k')
ax.set_ylabel('Log2 Normalizes Expression')
```

Output:

The boxplot in Figure 7.6 clearly shows that all of the quantiles are the same. That is why this technique of normalization is called quantile normalization.

For a microarray, data expression values of each gene are called features, and the samples are its instances. Generally, we put features as columns and instances as rows. Accordingly, we will interchange the rows and columns of the dataframe. This operation - named the transpose ".T" method of the dataframe - produces a transposed dataframe of itself.

TABLE 7.2

Quantile Normalized Values

	Control_1	Control_2	Control_3	Control_4	Treated_1	Treated_2	Treated_3	Treated_4
0	1690.3250	957.6500	1399.3875	1148.4500	1567.7625	1657.0000	5001.5625	1915.5625
1	736.6500	492.1250	231.0500	673.5125	426.2500	459.1875	356.2875	239.6375
2	884.4500	896.8250	530.5125	1360.4500	1106.2500	814.8250	1246.3750	935.1875
3	435.7000	678.8625	377.1500	616.9625	210.9375	284.3875	483.1625	286.4875
4	1416.9500	1365.9875	1340.7375	1956.2125	1469.4875	1924.3625	881.1000	1299.6375
...	...	...	...	...	...	...	...	...
16520	338827.9375	338827.9375	338827.9375	338827.9375	300324.4625	338827.9375	319862.6500	319862.6500
16521	300324.4625	284672.4500	319862.6500	319862.6500	272576.7750	308535.5750	300324.4625	272576.7750
16522	573472.2375	573472.2375	573472.2375	573472.2375	573472.2375	573472.2375	573472.2375	573472.2375
16523	458600.5000	458600.5000	458600.5000	458600.5000	458600.5000	458600.5000	458600.5000	458600.5000
16524	69.5000	73.7750	15.5000	63.9625	80.4125	72.6500	105.5750	24.3750

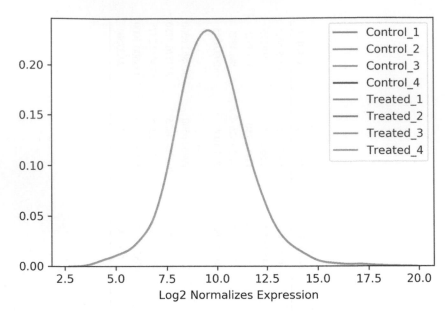

FIGURE 7.5   Quantile Normalized Values Distribution.

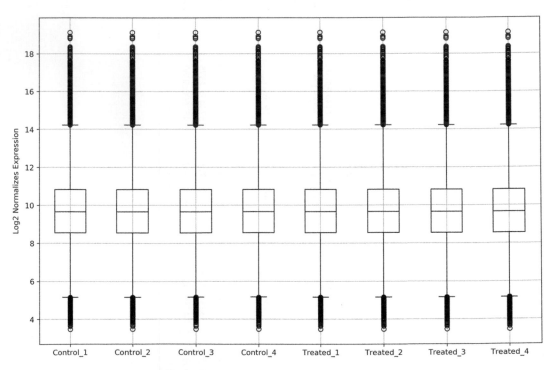

FIGURE 7.6   Boxplot for Quantile Normalize Values.

Code:

```
expression_transpose = log_norm_expression.T
expression_transpose.info()
expression_transpose.head()
```

TABLE 7.3

Transposed Dataframe, where Rows are Samples and Columns are Gene Expression Values

0	1	2	3	4	5	6	7	8	9
Control_1	10.723085	9.524836	9.788637	8.767191	10.468573	8.263621	9.025659	9.553869	6.403225
Control_2	9.903355	8.942881	9.808683	9.406976	10.415729	8.187723	9.523390	8.288001	8.408011
Control_3	10.450580	7.852061	9.051243	8.558995	10.388811	8.257977	8.407162	8.923959	9.059987
Control_4	10.165472	9.395561	10.409868	9.269039	10.933847	9.161478	10.654714	8.357442	8.972908
Treated_1	10.614491	8.735556	10.111462	7.720672	10.521097	9.742730	10.074610	8.486081	8.186548

Output:

```
<class 'pandas.core.frame.DataFrame'>
Index: 8 entries, Control_1 to Treated_4
Columns: 16525 entries, 0 to 16524
dtypes: float64(16525)
memory usage: 1.0+ MB
```

The dataset now contains eight rows representing the samples and 16,525 columns. This is an example of high-dimensional data, and this data set has 16,525 dimensions. We will use PCA to reduce its dimension to visualize the data points in 2D space. As we have discovered, the first step in PCA is to scale the data by subtracting the features from their means and then dividing this with standard deviation (Table 7.3).

Code:

```
expression_transpose_std = (expression_transpose-expression_transpose.mean())/
expression_transpose.std()
expression_transpose_std = expression_transpose_std.replace
    ([np.inf, -np.inf], np.nan)
expression_transpose_std = expression_transpose_std.dropna(axis=1)
```

The code above scales the values. Sometimes with real-world data, scaling may result in infinite values while dividing with standard deviations. These cases are rare, and we may have to exclude those values from our dataset. Therefore, the second and third statements are for converting the infinite values to "NaN" and dropping them using the ".dropna()" method. Finally, the dataset is ready for PCA implementation.

Code:

```
sklearn_pca = PCA(n_components=2)
print(expression_transpose_std.shape)
PCs = sklearn_pca.fit_transform(expression_transpose)
print(PCs.shape)
print(sklearn_pca.explained_variance_ratio_)
```

Output:

```
(8, 16523)
(8, 2)
[0.20351873 0.18757078]
```

After PCA implementation, the dataset dimensions have been reduced from 8 × 16,523 (i.e. two features are dropped as they produced infinite values while scaling) to 8 × 2, while preserving around 39% of variations within the dataset. Let us plot the data points in two dimensions:

Code:

```
expression_PCs = pd.DataFrame(PCs,columns=['PC1','PC2'])
expression_PC1 = expression_PCs.set_index(
    np.array((['Control']*4)+['Treated']*4))
expression_PC2 = expression_PCs.set_index(expression_transpose.index)
fig, axes = plt.subplots(figsize=(10,8))
sns.set_style("whitegrid")
```

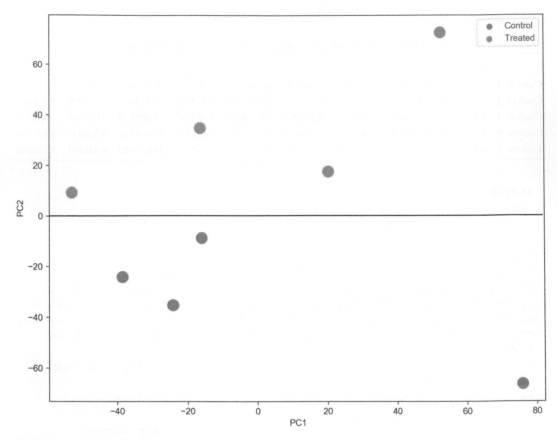

FIGURE 7.7   Plotting the Expression Data in Two Dimensions.

```
sns.scatterplot(x='PC1',y='PC2',data = expression_PC2,
                hue =expression_PC1.index, s=200)
```
Output:

The two-dimension plot with 39% variance (Figure 7.7) presents that the control and treated data can be separable through a line parallel to PC1 passing through the origin of PC2. This implies that the control groups and treated groups are different, and the data is suitable for observing the disparities among them to find differentially expressed genes. In many cases, PCA visualization will show groups of control and treated values placed distinctively next to each other, which also implies the data is applicable for further analysis.

Differential Expression Analysis

To discover the differentially expressed genes, the average values of control and treated groups for each gene are compared to see if the treated gene is upregulated or downregulated, with respect to the control gene. For statistical validation of this comparison, we use a hypothesis test called the student's t-test. The t-test is used for comparing means of samples. It provides a p-value, which is the estimated probability of rejecting the null hypothesis of a study question when that hypothesis is true. Therefore, a lower p-value implies that there is a significant difference between the expressions of a gene in two datasets. A p-value of less than 0.05 is statistically significant. It indicates strong evidence against the null hypothesis, because there is less than a 5% probability that the null is correct, and the results are random. Moreover, we reject the null hypothesis and accept the alternative hypothesis. After determining that the gene expression is significantly different between the sample sets, we will then calculate the change in expression by subtracting the log2 transformed values of the intensities of the control with the treated, or the Log2 Fold change. If the result is a positive value, then the gene is said to be upregulated. Consequently, if the result is a negative value, then it would be called downregulated. In executing the *t*-test, we will use a library of Python called "scipy" which popularly used for statistical and scientific calculations. The code is as follows:

Code:

```
from scipy import stats
control_mean = log_norm_expression.iloc[:,:4].mean(axis=1)
treated_mean = log_norm_expression.iloc[:,4:].mean(axis=1)
compair_means=pd.DataFrame(columns=['Control_Mean','Treated_mean'])
compair_means['Control_Mean']=control_mean
compair_means['Treated_mean']=treated_mean
compair_means['log2_fold_change'] = (compair_means['Treated_mean']-
                                    compair_means['Control_Mean'])
t_tset_values = pd.DataFrame(stats.ttest_ind
                (log_norm_expression.iloc[:,:4],log_norm_expression.iloc
[:,4:],axis=1)).T
t_tset_values.columns = ['Stastics','Pvalue']
compair_means   =   pd.concat([expression.iloc[:,0],compair_means,t_tset_
values],axis=1)
compair_means.head()
```

Output:

To achieve the table above, we keep the spot ids (i.e. the "ID_REF") columns so that we can combine these with the annotation table to assign the gene ids for the respective rows. In the next step, we will merge the annotation table with Table 7.4 with respect to the spot ids.

Code:

```
merged_table = pd.merge(annotation,compair_means,on='ID_REF')
merged_table.head()
```

Output:

Table 7.5 contains the gene details, their expression values for control and treated samples, $\log_2$ fold change, and the statistical values. The next step is to filter the rows with genes that are significantly up- or down-regulated in control and treated samples (p-value ≤ 0.05)

Code:

```
significant_table = merged_table[merged_table['Pvalue']<=0.05]
significant_table.info()
```

Output:

```
<class 'pandas.core.frame.DataFrame'>
Int64Index: 1251 entries, 23 to 16496
Data columns (total 9 columns):
 #     Column              Non-Null Count      Dtype
---    ------              --------------      -----
```

TABLE 7.4

Table of Compared Gene Expression

	ID_REF	**Control_Mean**	**Treated_mean**	**log2_fold_change**	**Statistics**	**_p_-Value**
0	1007_s_at	10.310623	11.125141	0.814518	1.891712	0.107397
1	121_at	8.928835	8.490026	−0.438809	−1.011076	0.351002
2	1316_at	9.764608	9.983611	0.219003	0.708891	0.504990
3	1552257_a_at	9.000550	8.237770	−0.762780	−2.384782	0.054412
4	1552264_a_at	10.551740	10.389579	−0.162161	−0.607118	0.566024

TABLE 7.5

Annotated Table with Gene IDs and Statistical Values

ID_REF	GB_ACC	Gene Symbol	ENTREZ_GENE_ID	Control_Mean	Treated_mean	log2_fold_change	Statistics	Pvalue
1007_s_at	U48705	DDR1 /// MIR4640	780 /// 100616237	10.310623	11.125141	0.814518	1.891712	0.107397
121_at	X69699	PAX8	7849	8.928835	8.490026	-0.438809	1.011076	0.351002
1316_at	X55005	THRA	7067	9.764608	9.983611	0.219003	0.708891	0.504990
1552257_a_at	NM_015140	TTLL12	23170	9.000550	8.237770	-0.762780	-2.384782	0.054412
1552264_a_at	NM_138957	MAPK1	5594	10.551740	10.389579	-0.162161	0.607118	0.56024

0	ID_REF	1251 non-null	object
1	GB_ACC	1250 non-null	object
2	Gene Symbol	1119 non-null	object
3	ENTREZ_GENE_ID	1100 non-null	object
4	Control_Mean	1251 non-null	float64
5	Treated_mean	1251 non-null	float64
6	log2_fold_change	1251 non-null	float64
7	Stastics	1251 non-null	float64
8	Pvalue	1251 non-null	float64

```
dtypes: float64(5), object(4)
memory usage: 97.7+ KB
```

We obtained 1,251 such genes that were significantly differentially expressed. Next, we place a cutoff on the $\log_2$ fold change values to produce stricter values. This cutoff depends upon the experiments and results and may vary between ± 1 and ± 2. In this step, we take ± 1.5 (i.e. more than 1.5 and less than -1.5) $\log_2$ fold change.

Code:

```
Upregulated_genes = significant_table[
    significant_table['log2_fold_change']>1.5]
Downregulated_genes = significant_table[
    significant_table['log2_fold_change']<-1.5]
print('Upregulated Genes')
Upregulated_genes.info()
print('\nDownregulated Genes')
Downregulated_genes.info()
```

Output:

```
Upregulated Genes
<class 'pandas.core.frame.DataFrame'>
Int64Index: 69 entries, 136 to 16178
Data columns (total 9 columns):
```

#	Column	Non-Null Count	Dtype
---	------	--------------	-----
0	ID_REF	69 non-null	object
1	GB_ACC	69 non-null	object
2	Gene Symbol	47 non-null	object
3	ENTREZ_GENE_ID	46 non-null	object
4	Control_Mean	69 non-null	float64
5	Treated_mean	69 non-null	float64
6	log2_fold_change	69 non-null	float64
7	Stastics	69 non-null	float64
8	Pvalue	69 non-null	float64

```
dtypes: float64(5), object(4)
memory usage: 5.4+ KB
Downregulated Genes
<class 'pandas.core.frame.DataFrame'>
Int64Index: 63 entries, 71 to 16262
Data columns (total 9 columns):
```

#	Column	Non-Null Count	Dtype
---	------	--------------	-----
0	ID_REF	63 non-null	object
1	GB_ACC	63 non-null	object
2	Gene Symbol	59 non-null	object

TABLE 7.6

Table of Differentially Expressed Genes

ID_REF	GB_ACC	Gene Symbol	ENTREZ_GENE_ID	Control_Mean	Treated_mean	log2_fold_change	Stastics	p-Value
1553253_at	NM_080863	ASB16	92591	7.767619	9.307180	1.539562	−3.317525	0.016054
1553954_at	BU682208	ALG14	199857	6.503933	8.086674	1.582741	−2.476600	0.048026
1554043_a_at	BC012528	NaN	NaN	6.857222	8.667420	1.810198	−2.667561	0.037147
1554772_at	BC036407	EFCAB13	124989	8.644272	10.226655	1.582383	−2.905322	0.027145
1555118_at	BC029869	ENTPD3	956	5.923432	7.722591	1.799159	−3.371086	0.015022

3	ENTREZ_GENE_ID	58 non-null	object
4	Control_Mean	63 non-null	float64
5	Treated_mean	63 non-null	float64
6	log2_fold_change	63 non-null	float64
7	Stastics	63 non-null	float64
8	Pvalue	63 non-null	float64

dtypes: float64(5), object(4)

memory usage: 4.9+ KB

We obtain 69 upregulated genes and 63 downregulated genes. Let us now combine the up- and down-regulated gene in a single table to obtain a consolidated table of differentially expressed genes (Table 7.6).

Code:

```
DEGs = pd.concat([Upregulated_genes,Downregulated_genes])
DEGs.head()
```

Output:

Cluster Map

Cluster maps are used to find co-expressed genes that may have genetic or physical interactions. We will use the log-transformed intensities and the GenBank accession id (i.e. the GB_ACC column) to build the cluster map. Let us merge the log transformed data and the GB_ACC column:

Code:

```
log_expression=pd.concat([expression.iloc[:,0],log_expression],axis=1)
table = pd.merge(DEGs.iloc[:,:3],log_expression,on='ID_REF')
index = table.iloc[:,1]
table = table.iloc[:,3:].set_index(index)
table.head()
```

Output:

The cluster map is built using Seaborn's ".clustermap" method, which uses a matrix of numbers to build a cluster map (Table 7.7).

Code:

```
sns.clustermap(table,cmap='coolwarm')
```

Output:

In Figure 7.8, we observe the hierarchical clustering of differentially expressed genes, more specifically, the shades are divided into two groups horizontally - control and treated. This also divided into two clusters of upregulated and downregulated genes. Orange indicates increased gene expression levels; blue indicates decreased levels. Genes with similar expression values are placed nearer and, hence, may be co-expressed.

TABLE 7.7

Data for Building a Cluster Map

GB_ACC	Control_1	Control_2	Control_3	Control_4	Treated_1	Treated_2	Treated_3	Treated_4
NM_080863	8.163398	7.315602	7.305606	8.657140	9.642052	8.896332	9.869440	8.719047
BU682208	7.779391	7.172927	3.944858	6.507795	7.837943	8.327777	8.751544	7.235536
BC012528	8.587215	7.062856	6.277985	4.336283	8.104861	8.756223	9.707532	7.955940
BC036407	8.820179	9.316734	9.170426	7.210428	9.704077	10.421329	11.122569	9.643676
BC029869	7.830990	4.209453	6.137504	4.078951	7.001127	7.326429	8.431289	7.835419

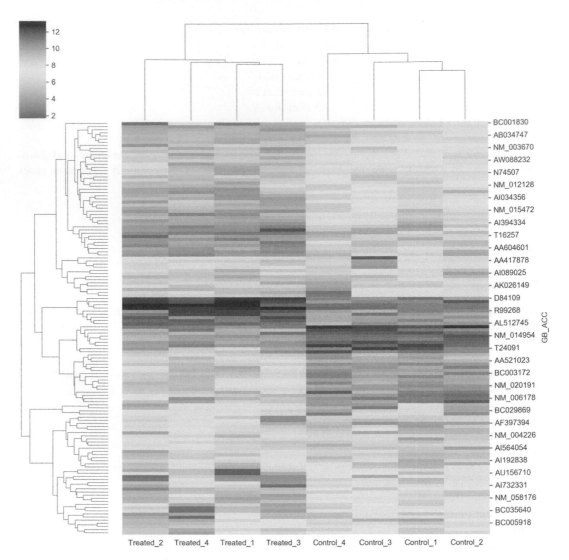

FIGURE 7.8 Cluster Map of DEGs.

Gene Enrichment Analysis

Gene enrichment analysis (GEA) is the biological interpretation of the results. We perform GEA to decipher the biological processes and molecular pathways being affected in the case of treated samples. For GEA, we use the hypergeometric test statistics ($p \leq 0.05$) to filter out random matches. In executing GEA, we use a Python library called "gseapy".

Code:

```
import gseapy as gp
enr = gp.enrichr(gene_list=gene_list,
                 gene_sets=['GO_Biological_Process_2017b'],
                 organism='Human'
                 description='test_name',
                 outdir='test/enrichr_kegg',
                 cutoff=0.5 # test dataset, use lower value from range(0,1))
enr.results[['Gene_set','Term','Overlap','P-value','Genes']].head()
```

TABLE 7.8

Gene Enrichment Analysis Results Using Gseapy

	Gene_set	Term	Overlap	*p*-Value	Genes
0	GO_Biological_Process_2017b	Modulation by the organism of defense-related calcium ion flux in other organism involved in symbiotic interaction (GO:0052301)	2/8	0.000763	RIT2;MAPT
1	GO_Biological_Process_2017b	Modulation by the organism of defense-related calcium-dependent protein kinase pathway in other organism involved in symbiotic interaction (GO:0052307)	2/8	0.000763	RIT2;MAPT
2	GO_Biological_Process_2017b	Ganglioside biosynthetic process (GO:0001574)	3/34	0.000768	B4GALNT1; SPTSSA; ST6GALNAC5
3	GO_Biological_Process_2017b	Regulation of calcium-mediated signaling (GO:0050848)	2/9	0.000978	RIT2;MAPT
4	GO_Biological_Process_2017b	Regulation of CAMKK-AMPK signaling cascade (GO:1905289)	2/9	0.000978	RIT2;MAPT

Output:

The gseapy library has a method called "enrichr()" – this primarily uses a list of differentially expressed genes, the organism, the list database, or the library we prefer to search against, the cutoff p-value, and the directory where we want to save the results. The list of databases can be found on "https://amp.pharm.mssm.edu/Enrichr/#stats" or by applying a method called the "gp.get_library_name()" method. The top five biological processes affected due to our DEGs are shown in Table 7.8:

The procedure discussed above has shown the step-by-step guide of microarray data analysis. More sensitive techniques like RNAseq have been developed where expressions are inferred quantitatively using fragment counts compared to fluorescent intensity in microarray data.

SNP Analysis

Genetic variants are key drivers of evolution. The most common form of genetic variation is single nucleotide polymorphism (SNPs). These SNPs are also called biomarkers and have been associated with various phenotypes, including disease susceptibility and drug response or pharmacogenomics. Researchers identify SNPs and their associations with phenotype by comparing the genomes of large populations. There are various phenotype-genotype association databases like SNPedia, ClinVar, and PhenGen, among others. SNPedia is a curated form of scientific studies for SNP association with various phenotypes. It is publicly available and is also used by industries. ClinVar and PhenGen are hosted by NCBI, where ClinVar contains rather specialized association data where SNPs associated with clinical implications are stored, and PhenGen accumulates data from various association studies.

Variants can be determined to form genomic sequences by aligning these to a reference genome, where variants are well annotated. This process is called variant calling, and this saves a file where variants of the individual are determined with the genotypes. An individual may have millions of SNPs, and, therefore, the SNP data can safely be categorized as big data. Handling this amount of data requires knowledge of programming languages for conducting customized studies. In the following example, we will use a publicly available genotype data form called Kaggle "Zeeshan-ul-hassan Usmani, Genome Phenotype SNPS Raw Data File by 23andMe, Kaggle Dataset Repository, Jan 25, 2017". This data constitutes the SNP genotype data of an individual. We will map this data with drug response data from a pharmacogenomics database called PharmGKB (https://www.pharmgkb.org/). Let us load the geno-type data, which is available in the supplementary file with this chapter and is labeled as "genotype.csv".

Code:

```
genotype = pd.read_csv('genotype.csv')
print(genotype.head())
genotype = genotype.rename(columns={"# rsid":"ID"})
print(genotype.head())
```

Output:

#	rsid	chromosome	position	genotype
0	rs12564807	1	734462	AA
1	rs3131972	1	752721	AG
2	rs148828841	1	760998	AC
3	rs12124819	1	776546	AA
4	rs115093905	1	787173	GG

	ID	chromosome	position	genotype
0	rs12564807	1	734462	AA
1	rs3131972	1	752721	AG
2	rs148828841	1	760998	AC
3	rs12124819	1	776546	AA
4	rs115093905	1	787173	GG

This data contains an ID field "# rsid"; these are unique IDs for SNPs. Next, we have identified the chromosome number and genomic location of the SNPs as the "position" column. Lastly, we have the genotype for the SNP. We changed the name of "# rsid" to "ID".

Let us study how many SNPs are present per chromosome. The Pandas ".value_count()" function can do this. It is used on a dataframe column, which counts the occurrence of unique values in a column.

Code:

```
genotype["chromosome"].value_counts()
```

Output:

```
1     47742
2     46815
6     40965
3     39186
5     34900
4     34386
7     33552
8     30651
11    29883
10    29592
12    29068
9     27010
13    21882
16    19680
X     19588
17    19364
14    19002
15    18660
20    14730
18    14696
19    13948
22     9307
21     8571
```

```
MT    3287
Y     2129
18    1950
Name: chromosome, dtype: int64
```

Next, we will download the genotype-phenotype association data from PharmGKB (https://www. pharmgkb.org/downloads). This data contains the association of drug response with genetic variants.

Code:

```
phenotype = pd.read_csv('phenotype.tsv',sep='\t')
phenotype.info()
```

Output:

```
<class 'pandas.core.frame.DataFrame'>
RangeIndex: 4561 entries, 0 to 4560
Data columns (total 16 columns):
```

#	Column	Non-Null Count	Dtype
0	Unnamed: 0	4561 non-null	int64
1	Clinical Annotation ID	4561 non-null	int64
2	Location	4561 non-null	object
3	Gene	4334 non-null	object
4	Level of Evidence	4561 non-null	object
5	Clinical Annotation Types	4538 non-null	object
6	Genotype-Phenotype IDs	4561 non-null	object
7	Annotation Text	4561 non-null	object
8	Variant Annotations IDs	4561 non-null	object
9	Variant Annotations	4561 non-null	object
10	PMIDs	4561 non-null	object
11	Evidence Count	4561 non-null	int64
12	Related Chemicals	4561 non-null	object
13	Related Diseases	3633 non-null	object
14	Biogeographical Groups	4553 non-null	object
15	Chromosome	4076 non-null	object

```
dtypes: int64(3), object(13)
memory usage: 570.2+ KB
```

The data consists of 15 columns for annotations, IDs, chemicals, diseases, and more. Out of all of these, we will keep four columns, namely, location (which essentially contains the SNP rsids that will be required to merge with the genotype table), the annotation text (which has information about the drug response), related chemicals, and related diseases to find the associated drugs and disease (Table 7.9).

Code:

```
pd.set_option('display.max_colwidth', 200)
phenotype = phenotype.rename(columns = {"Location":"ID"})
phenotype=pd.concat([phenotype['ID'],phenotype['Annotation Text'],
                     phenotype['Related Chemicals'],
                     phenotype['Related Diseases']],
                     axis=1)
phenotype.head()
```

Output:

We will also retain only two columns for the genotype table - namely, the IDs and the genotype version or the allele (Table 7.10).

Code:

```
genotype = pd.concat([genotype['ID'],genotype['genotype']], axis=1)
genotype.head()
```

TABLE 7.9

Phenotype Table Containing Drug Responses Associated with Genotypes from PharmGKB

ID	Annotation Text	Related Chemicals	Related Diseases
rs75527207	AA: Ivacaftor is indicated in cystic fibrosis patients with the AA genotype (two copies of the CFTR G551D variant). FDA-approved drug labeling information indicates the use of ivacaftor in cystic fibro…	Ivacaftor (PA165950341)	Cystic fibrosis (PA443829)
rs4149056	CC: Patients with the CC genotype and Precursor Cell Lymphoblastic Leukemia-Lymphoma may need a decreased dose of mercaptopurine, or methotrexate, as compared to children with the TT genotype. Other…	Mercaptopurine (PA450379); methotrexate (PA450428)	Precursor cell lymphoblastic leukemia-lymphoma (PA446155)
rs141033578	CC: Patients with the CC genotype (do not have a copy of the CFTR S977F variant) and cystic fibrosis have an unknown response to ivacaftor treatment, as the response may depend on the presence of other…	Ivacaftor (PA165950341)	Cystic fibrosis (PA443829)
rs78769542	AA: Patients with the AA genotype (two copies of the CFTR R1070Q variant) and cystic fibrosis may respond to ivacaftor treatment. FDA-approved drug labeling information and CPIC guidelines indicate…	Ivacaftor (PA165950341)	Cystic fibrosis (PA443829)
rs1799971	AA: Patients with AA genotype may have an increased likelihood of smoking cessation when treated with nicotine replacement therapy (transdermal nicotine patch) as compared to patients with the AG a…	Drugs used in nicotine dependence (PA164712720); nicotine (PA450626)	Tobacco use disorder (PA445876)

Output:

Now, let us merge both the genotype (Table 7.10) and the phenotype (Table 7.9) tables and study the results.

Code:
```
pheno_table = pd.merge(genotype,phenotype,on='ID')
pheno_table.info()
```
Output:
```
<class 'pandas.core.frame.DataFrame'>
Int64Index: 2331 entries, 0 to 2330
```

TABLE 7.10

Table Containing Genotype IDs and Versions of the Alleles

ID	Genotype
rs12564807	AA
rs3131972	AG
rs148828841	AC
rs12124819	AA
rs115093905	GG

TABLE 7.11

Associations Grouped by Chemicals

	ID				Genotype				Annotation Text			
	count	Unique	Top	Freq.	count	Unique	Top	Freq.	Count	Unique	Top	Freq.
Related chemicals												
3,4-Methylenedioxymethamphetamine (PA131887008)	2	2	rs2242446	1	2	2	GT	1	2	2	CC: Individuals with the CC genotype who are exposed to (3,4-methylenedioxymethamphetamine) MDMA may have an increased response, specifically an increased heart rate, as compared to patients with t...	1
ABT-751 (PA166104276)	1	1	rs6755571	1	1	1	CC	1	1	1	AA: Patients with AA genotype were not studied. However, patients with the AC genotype and neoplasms may have a decreased plasma predose concentration as compared to patients with the CC genotype....	1
Ace inhibitors, plain (PA164712308)	10	10	rs495828	1	10	6	CT	2	10	10	AA: Patients with the AA genotype who are treated with ACE inhibitors may have an increased risk for cough as compared to patients with the GG genotype. Other genetic and clinical factors may also...	1
Ace inhibitors, plain (PA164712308);angiotensin II antagonists (PA164712372); beta blocking agents (PA164712535); digoxin (PA449319); diuretics (PA151249535); spironolactone (PA451483)	2	2	rs1801253	1	2	2	GG	1	2	2	AA: Patients with the AA genotype and heart failure may have decreased emergency department visits and hospital utilization when treated with cardiovascular drugs as compared to patients with the A...	1
Alkylating agents (PA164712331)	1	1	rs712829	1	1	1	--	1	1	1	GG: Cancer cells with the GG genotype may be more sensitive to Alkylating agents than are cells with genotype GT or TT. Other genetic and clinical factors may also influence tumor response to Alkyl...	1

TABLE 7.12

Phenotype-Genotype Association for Drug Warfarin

	ID	Genotype	Annotation Text	Related Chemicals	Related Diseases
123	rs2501873	CT	CC: Patients with the CC genotype and any allele of rs3212198 who are treated with warfarin may require a lower dose as compared to patients with the TT genotype and rs3212198 T allele. The variant...	Warfarin (PA451906)	NaN
175	rs1877724	CT	CC: Patients with the CC genotype may require an increased dose of warfarin as compared to patients with the TT genotype. Other genetic and clinical factors may also influence a patient's dose of w...	Warfarin (PA451906)	NaN
206	rs679899	AA	AA: Patients with the AA genotype and receiving warfarin following cardiac valve replacement may have a decreased risk of bleeding at therapeutic INR as compared to patients with the AG or GG genot...	Warfarin (PA451906)	Heart valve replacement (PA166123431); hemorrhage (PA444417)
208	rs1367117	GG	AA: Patients with the AA genotype and receiving warfarin following cardiac valve replacement may have a decreased risk of bleeding at therapeutic INR as compared to patients with the AG or GG genot...	Warfarin (PA451906)	Heart valve replacement (PA166123431); hemorrhage (PA444417)
223	rs11676382	CC	CC: Patients with the CC genotype may have decreased international normalized ratio variability (INR-var) when treated with warfarin as compared to patients with genotype GG or CG in African-Americ...	Warfarin (PA451906)	NaN
224	rs11676382	CC	CC: Patients with the CC genotype may have increased time in therapeutic range when treated with warfarin as compared to patients with genotype GG or CG in African-Americans after the warfarin dose...	Warfarin (PA451906)	NaN
225	rs11676382	CC	CC: Patients with the CC genotype may need an increased dose of warfarin as compared to patients with the CG and GG genotypes, however, this has been contradicted in some studies. Other clinical and...	Warfarin (PA451906)	NaN
226	rs2592551	GG	AA: Patients with the AA genotype and atrial fibrillation may require a higher dose of warfarin as compared to patients with the GG genotype. Other genetic and clinical factors, such as variations...	Warfarin (PA451906)	Atrial fibrillation (PA443459)
227	rs12714145	CC	CC: Genotype CC may be associated with a decreased dose of warfarin as compared to genotype TT, although this is contradicted in most studies.	Warfarin (PA451906)	NaN

(Continued)

TABLE 7.12 (*Continued*)

ID		Genotype	Annotation Text	Related Chemicals	Related Diseases
310	rs887829	CT	Other genetic and clinical factors may influence a patie... CC: Patients with the CC genotype and heart valve replacement may require a lower stable dose of warfarin compared to patients with the CT and TT genotypes, although this is contradicted in one stu...	Warfarin (PA451906)	Heart valve replacement (PA166123431)
860	rs41301394	CT	CC: Patients with the CC genotype may require a decreased dose of warfarin as compared to patients with the CT or TT genotype. Other genetic and clinical factors may also impact the dose of warfarin....	Warfarin (PA451906)	NaN
875	rs1045642	GG	AA: Patients with the AA genotype may require an increased dose of warfarin as compared to patients with the AG or GG genotypes, although this is contradicted in one study which found the opposite...	Warfarin (PA451906)	NaN
1153	rs339097	AA	AA: Patients with the AA genotype who are treated with warfarin may require a lower maintenance dose as compared to patients with the AG or GG genotype, although this is contradicted in one study....	Warfarin (PA451906)	NaN
1163	rs2645400	GT	GG: Patients with the GG genotype may require decreased doses of warfarin as compared to patients with the TT genotype. Other genetic and clinical factors may also influence warfarin dose, such as...	Warfarin (PA451906)	NaN
1191	rs4379440	GG	GG: Patients with the GG genotype may have increased time in therapeutic range (TTR) when treated with warfarin as compared to patients with genotype TT. Other genetic and clinical factors may infl...	Warfarin (PA451906)	NaN

```
Data columns (total 5 columns):
 #            Column              Non-Null Count        Dtype
---           ------              --------------        -----
 0            ID                  2331 non-null         object
 1            genotype            2331 non-null         object
 2            Annotation Text     2331 non-null         object
 3            Related Chemicals   2331 non-null         object
 4            Related Diseases    1873 non-null         object
dtypes: object(5)
memory usage: 109.3+ KB
```

After merging both tables, we are able to obtain 2,331 associations of phenotypes, chemicals, and diseases. To analyze this further, we are aware that drugs are metabolized by various enzymes, and single enzymes can have more than one SNP - which may affect the metabolism of the drugs. Therefore, it is conclusive that one drug or chemical can have more than one SNP association for its metabolism. Now, we group the SNPs based on their association with chemicals. For this purpose, we use will use the Pandas ".groupby()" function:

Code:

```
pheno_table.groupby('Related Chemicals').describe().head()
```

Output:

The first or count column provides the number of associations for each drug.

Next, we will convert Table 7.11 into a dictionary, where keys will be "Related Chemicals", and the items will be the dataframe of associated genotype-phenotype association. Therefore, we can use the keys (i.e. the drugs) to determine all of the linked results.

Code:

```
drugs_phenotype = {k: v for k, v in pheno_table.groupby('Related Chemicals')}
drugs_phenotype['warfarin (PA451906)'][:15]
```

Output:

Warfarin is an anticlotting agent that is used for the prevention of stroke in individuals who have atrial fibrillation and artificial heart valves. From Table 7.12, we can estimate the dose of warfarin while studying the alleles in the genotype. Similarly, we can group the data according to the diseases and confirm the associations for that disease with genotype.

Exercise

1. The "practice_expression.csv" file in the supplementary file contains the mRNA expression profile of normal versus SARS-infected patients retrieved from GEO (ID: GSE1739), with its corresponding annotations in "practice_annotation.csv". Determine the significantly up- and down-regulated mRNAs.

2. Next, perform gene enrichment analysis to discover the biological processes affected by the SARS infection.

3. Draw a cluster map to determine the associated genes.

4. The "practice_Phenotype.tsv" file contains the general traits associated with "rsid", retrieved from PheGen1 (https://www.ncbi.nlm.nih.gov/gap/phegeni). Use the genotype data used in this chapter to match the rsids with "practice_Phenotype.tsv".

8

Machine Learning and Linear Regression

Introduction to Machine Learning and Its Applications in Biology

The art of enabling machines to form rules and find trends from data without explicitly programming them is known as Machine Learning. In Machine Learning, we allow computers to gather intelligence from the data and adapt to their situations based on their experience. The applications of Machine Learning are tremendous, because Machine Learning has the potential to affect every domain of our daily lives.

Throughout modern history, humans have been using machines to reduce efforts and increase efficiency in doing work, and Machine Learning is an extension of this. Extensive research is taking place all over the world so that we can use these latest methods to prepare ourselves for the unfamiliar challenges of the future. Companies have started working with bots and web crawlers. They have developed systems to an extent where these have started to learn on their own. The capabilities of Machine Learning are endless, and the world, as we see it today, will never be the same.

Machine Learning is required to extract meaningful information from big datasets. It can also be used to alleviate the burden of solving many biological problems. Today, we have numerous examples of applied Machine Learning - from protein structure prediction, image recognition, drug molecule development, drug repurposing, protein-protein interaction, finding SNPs in genomic sequences, cancer detection, solving problems in system biology, biological text mining, and many more. Protein structure prediction algorithms before Machine Learning had an accuracy rate of only around 70%. However, Machine Learning has pushed the boundaries in this field as well, causing the accuracy rate to surge at a much better figure of 85%. With the advancement in high-throughput technologies in biology as well as the increasing number of publications, data is growing in size more rapidly than it ever has. Machine Learning systems can now use this data to perform text-mining for research purposes. These systems work similarly to how the human brain does but have the computational power that far surpasses our capabilities. Outputs from such a system can also be used in identifying microscopic cellular images and performing various medical studies.

Types of Machine Learning Systems

There are two main types of Machine Learning methods: supervised and unsupervised.

Supervised Learning

Supervised learning is when our model becomes trained on a pre-labeled dataset and develops systems to predict outcomes for unforeseen data. The data used to train the system in supervised learning contains both input and output values.

This type of Machine Learning is then further classified into classification and regression. In classification, output has discrete values. Our system will have to classify these values into their respective groups correctly. An example is classifying SNPs into the functional or benign class, or predicting the day to be hot or cold. In regression, the output contains continuous values, so the system has to predict an output value close to the actual value. An example is the prediction of temperature, stock prices, and more.

Unsupervised Learning

Unlike supervised learning, unsupervised learning is used for clustering or inferring patterns, trends, or relationships within the dataset, without any prior reference or knowledge about the labeled data.

Unsupervised learning is further classified into clustering and anomaly detection. Clustering is the most popular technique in unsupervised Machine Learning. In clustering, the model groups the data into various clusters that have comparable parameters. We have already learned about an example of clustering while building a cluster map part of microarray analysis projects, where genes and samples were clustered into groups (chapter 7). On the other hand, anomaly detection functions by recognizing anything that differs from a general trend. For example, if we continuously see white cars on the road, then when a red car comes along, it will surely be a notable event. This is how anomaly detection systems perform their function. This method finds its applications in detecting bank frauds, human errors during data entry, and many more.

Figures 8.1 and 8.2 pictorially depict the differences between supervised and unsupervised learning algorithms:

Other Types of Machine Learning

Semisupervised Learning: This approach is used when we have a small dataset of known labeled data and massive unknown data. These models initially take the opportunity to study some trends from the known labels and then try to discover other hidden patterns from unlabeled data.

Reinforcement Learning: In this approach, an agent interacts with its surrounding environment using the actions for attending a user-defined goal and is trained using a reward or punishment method. Reinforcement learning has applications in self-driving cars in which reaching the end location is the user-defined goal, and driving skills like turning, accelerating, and stopping represent the actions. Each wrong action concludes with punishment or error, and the correct steps are accompanied by rewards.

It can be a daunting task to select which approach one needs to use to attain the optimal results. There is a multitude of aspects to be considered. Different problems require different solutions, and there is no single method that fits all. We must study the problem at hand in detail to understand which method will be best suited for that specific application. We should evaluate the information given to us, define our goals, review available algorithms, and study successful examples of similar problems in the past to be able to identify the Machine Learning method that is best suited for the mentioned problem.

Evaluation of Models

The best way we can evaluate the performance of our model is by not using all of the data while training. Usually, we divide the complete set of data into two parts - 80% of the information is used to train the machine, and 20% of the data is used to test how correctly the model can predict. A trained model must not be exposed to the test data during training. This ensures that any predictions made on the test dataset are indicative of the true performance of the model in general. This is the step where the efficiency of our system is determined. To check the accuracy of the model, we compare the predicted values with actual values in a confusion matrix (Figure 8.3).

Using the values of the confusion matrix, we can discover various mathematical measures for the evaluation of a model, such as accuracy, precision, recall, etc.

Accuracy

Accuracy is the rate of correct prediction for a model or the number of values correctly predicted divided by the total number of instances in the test set.

$$\text{Accuracy} = \{TP + TN\}/\{TP + FP + TN + FN\}$$

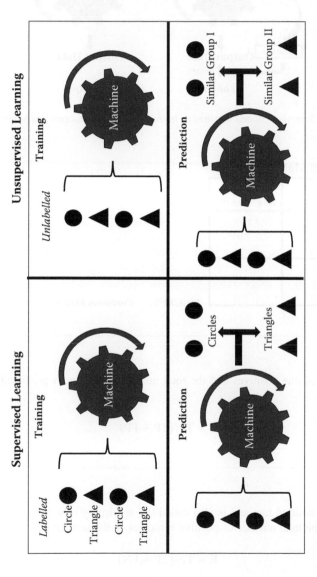

FIGURE 8.1 Supervised versus Unsupervised Learning. During Training, Supervised Models Know the Labels, While Unsupervised Learning Does Not Have Data Labels.

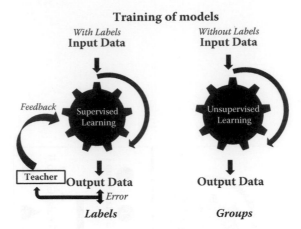

FIGURE 8.2 Traning of Supervised Models Required an Instructor, Whereas Unsupervised Models Learn on Their Own.

Confusion Matrix

	Actually Positive (1)	Actually Negative (0)
Predicted Positive (1)	True Positives (TPs)	False Positives (FPs)
Predicted Negative (0)	False Negatives (FNs)	True Negatives (TNs)

FIGURE 8.3 Confusion Matrix.

Precision

Precision is the ratio of true positives and the total number of instances predicted as positive by the model.

$$P = TP/\{TP + FP\}$$

Recall

The recall is the true positive rate and is also called the sensitivity of a model or the number of true positives divided by the total number of positive instances in the test dataset.

$$R = TP/\{TP + FN\}$$

F1 Score

Precision and recall are always mentioned together, such as precision at a recall level, or are measured in a single mathematical value called the F1 score. This is the harmonic mean of precision and recall.

$$F1\ Score = 2*((Precision*Recall)/(Precision + Recall))$$

Receiver Operating Characteristics (ROC) Curve

The Receiver Operating Characteristics (ROC) Curve is another method for the evaluation of the classifiers. It is a plot between the true positive rate (TPR) or the sensitivity and false positive rate (FPR) (1-Specificity) (Figure 8.4).

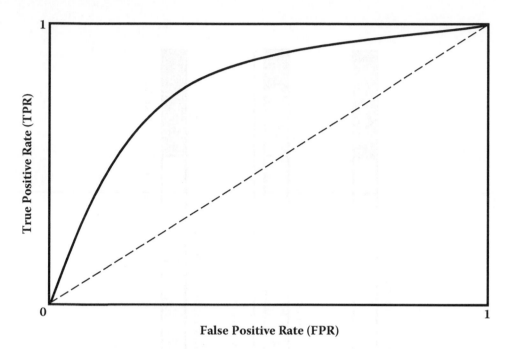

FIGURE 8.4 ROC Curve.

The solid curve is the change of the true positive rate with the false positive rate. The bigger the area is under the curve, the more the accuracy is of the model.

Evaluation metrics for regression models are fairly complex. Regression deals with a continuous dataset, which means advanced metrics are required for their assessment. Common metrics for regression model evaluation are variance - specifically, mean squared error and the R squared coefficient.

Cross-Validation

While performing any Machine Learning task, we usually divide our dataset randomly into two sets, as discussed above (i.e. the training and the test set). However, there is a limitation to this technique. If we sacrifice around 20% of our data for evaluating our system, then our model may not be able to train properly for the real-world challenges. Some valuable information might get lost when we split the dataset. This problem worsens in the case of a small dataset. To solve this problem, we use a technique called cross-validation. In this technique, we split the dataset into various subsets and then train and evaluate our model with various combinations of these subsets (Figure 8.5). These methods include the K-fold validation and the Leave-One-Out Cross-Validation (LOOCV), among others.

Optimization of Models

The optimization of models is a crucial step to improve the accuracy of the predicted results. Machine Learning models are trained on complex datasets; therefore, they have various hyperparameters that we

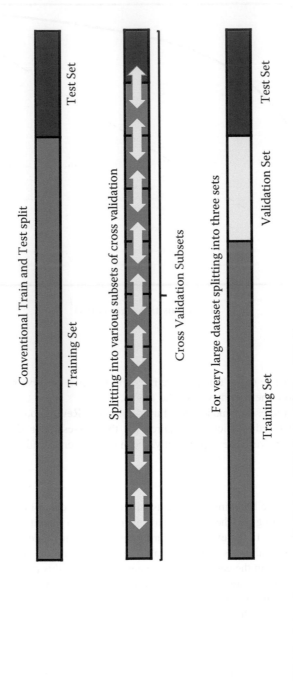

FIGURE 8.5 Various Arrangements for Training, Validating, and Testing of Models.

will learn in subsequent chapters. The selection of the right model and parameter values are key findings of finetuning a Machine Learning model. Generally, models are iteratively trained and evaluated using various combinations of parameters. By improving our model regularly and testing the accuracy of the model on a continuous basis with various hyperparameter combinations, we can ensure that the performance can be further optimized.

Grid Search

Grid searching is a method of scanning the space of all hyperparameter combinations for a model to discover the best combination of hyperparameters. Depending on the type of model utilized, certain hyperparameters are necessary. Grid searching can be applied to any Machine Learning model to calculate the best hyperparameters to use for that system. It is significant to note that grid searching is computationally intensive and may take a long time to run on a model. It iterates through every hyperparameter combination possible and stores a model for each combination.

Randomized Search

James Bergstra and Yoshua Bengio (2012) proposed the idea of random searching of hyperparameters in a system. This type of search is completely different from the grid approach. Instead of sweeping through every possible combination in hyperparameter space, a randomized search only selects a few sample points from the distribution and performs the calculations on those points.

Ensemble Methods

This is a combination of procedures in which various models are used together to form a significantly improved version. This allows us to acquire better results as compared to individual models. The two most widely used ensemble methods are averaging and voting. These are not complex and can easily be implemented to increase the accuracy of the system.

Averaging is performed while dealing with continuous data ranges, whereas voting is performed when dealing with a discrete classification of data. Initially, a number of different models are trained by different subsets of training data. Next, their results are combined. Therefore, instead of creating only one system and then assuming it to be the most accurate, ensemble methods adopt several systems and average these to produce one final product.

Challenges in Machine Learning Projects

Machine Learning gives us the ability to make more informed and data-driven decisions that are faster than conventional approaches. However, as with any other method, the Machine Learning process presents its own set of challenges.

Challenges with Data

a. Inadequate Training Data
   Having sufficient data for training in a system is extremely crucial for success in any Machine Learning project. Many of the Machine Learning systems fail miserably due to the lack of data. Having inadequate data means that the system is unable to understand the trends properly, and this massively compromises the efficiency of that system. This is the reason why data collection has become an integral part of the Machine Learning process. The amount of data needed depends both on the complexity of a problem and the algorithm selected.

b. Non-Representative Training Data

In order to generalize the model, it is of key importance that the training data is an accurate representation of the population under consideration. Each time a new subset is derived from the population, it is crucial that the subset must depict the complete picture of that dataset.

If the sample is too small, then we will gain sampling noise - which is classified as non-representative data. On the other hand, even large samples can be non-representative if the sampling method is flawed. This is known as sampling bias. It is important to note that if we reduce sampling bias, then the variance increases. In which case, the model is not able to generalize the unseen data; while, if the variance is reduced, the bias is boosted. This phenomenon, which is also known as the bias-variance trade-off, is the process of finding the right balance between both of these parameters.

Quality of Data

Poor data quality is a huge challenge in Machine Learning systems. These systems require high-quality data to avoid a situation in which they can fail either of the training and testing phases. Machine Learning is a data-intensive technique. That is why the quality of the data used in any Machine Learning system has a huge effect on its development. Because of this, small errors in the training data can lead to large-scale errors in the system's output. Increasingly complex problems require not just massive amounts of data, but also diverse data. It must be comprehensive as well as of good quality.

Inappropriate Features

Features that are used for training the system affect the efficiency in which we achieve the output. Irrelevant features can undermine any system and can also increase the cost of training.

Feature selection should be a part of the most basic stage while designing a Machine Learning project. This becomes even more important when one has a number of features. There is no need to use every feature available while developing the algorithm. We have to optimize our algorithm by feeding in only certain features that are considered most important and independent. Sometimes, less is better (i.e. we can use dimensionality reduction while keeping the variance), like in the case of PCA where we can use the principle components as features. We can also utilize various methodologies and techniques to select a subset of the feature space to help our models perform better, because more features bring more noise.

Until this point, we have discussed the challenges and the limitations of Machine Learning due to the unavailability of useful data. With the advancements in Machine Learning, data has become a new form of gold. Let us see the limitations of these models due to hyperparameters.

Overfitting and Underfitting the Training Data

Overfitting refers to a system that models the training data too well. This happens when a model learns unnecessary details and gains noise in the training data. This negatively impacts the performance of the system on test data.

Underfitting occurs when a system is not able to capture relationships between features and output variables precisely. This affects the ability of the algorithm to decipher the underlying trend of the data. Underfitting is a strong indicator that the algorithm not suitable for the dataset under consideration. Underfitting is common in cases where there is limited data available for training the system. Fewer data points result in a non-accurate training of the model which then fails to perform efficiently on the test dataset. Increasing the amount of data could be one way to ensure that the algorithm has enough information in which it can detect the general trends and patterns.

At this point, let us proceed towards the next section on Linear Regression to study the potential of Machine Learning. Here is an introduction to Scikit-Learn:

Before we go to different Machine Learning algorithms and gain hands-on experience in implementing those in Python, we will learn about a library called Scikit-Learn. Scikit-Learn or sklearn is one of Python's central Machine Learning packages. It has all of the modules required to carry out basic

Machine Learning tasks. All of the processing and estimator modules in this kit have a uniform flow for users, which are very user-friendly in terms of the execution of Machine Learning algorithms - such as classification or prediction with just a few lines of code. The models have dependencies on Pandas, NumPy, and SciPy packages in managing a diverse variety of datatypes. The package is well documented and includes references and tutorials for beginners.

As previously discussed, sklearn has a uniform workflow for estimator models. An estimator can be in the form of a data-processing pipeline like scaling the data and imputing missing data, among others, which transforms data or it can also be in the form of a Machine Learning algorithm execution. Figure 8.6 shows a general outline of the process flow of the sklearn module.

The most common flow of sklearn modules are:

1. The first step is to initiate a class of estimators with default or custom parameters.
2. Next is to pass the data to the estimator instance using the fit method. The data can be a Pandas dataframe, NumPy's ndarray, a Python nested list, etc. If the estimator is a Machine Learning algorithm, then the fit method trains the model on the given data.
3. If the estimator is a data processing module, then the transform method is used to process the data. We also have a fit_transform method for executing both fitting and transforming, thus combining steps 2 and 3.
4. In the case of a Machine Learning algorithm, the estimator has the predict method to evaluate the test data or predict any new instance.

Scikit-Learn is mostly used for the implementation of all of the general Machine Learning algorithms except artificial neural networks and deep learning. For deep learning, Python has packages like TensorFlow, PyTorch, and Theano, among others. In this book, we will study TensorFlow for artificial neural networks and discuss it in Chapter 14. We will be utilizing sklearn for linear regression, logistic regression, k-nearest neighbors, decision trees and random forests, support vector machines, and k-means clustering. In this chapter, we will learn how to carry out the linear regression using sklearn.

Linear Regression

Linear regression is a supervised learning algorithm that is used for the prediction of continuous variables - such as sales, age, product price, salary, and weather predictions, etc. Linear regression is represented by a linear equation in which one variable is independent (x), and the second one is the dependent variable or the output (y) (Smith et al., 1988).

If the input and output values are linearly related, then a line can be drawn with the data points, as shown in Figure 8.7. This will describe the relationship between the variables. In a simple linear regression model with a single x and a single y, the equation would be as follows:

$$y = mx + b$$

This illustrates the equation of the line - where "x" is the independent variable; "y" is the dependent variable; "m" is the slope; "b" is the y-intercept. The task of linear regression is to find the optimal values of "m" and "b" based on "x" and "y" data points so that the average distance between the points and the line is at the minimum. For a line, the distance between the points and the lines are called errors, and each point has its own error. When these errors are squared and summed, the value thus obtained is called the sum squared error, and the mean of this sum is called the mean square error. This mean square error is the cumulative error among all of the data points and the line.

As we can observe in Figure 8.8a, there can be many values for "m" and "b", regression finds the optimal line iteratively by fitting multiple lines and then selecting the line that has the least mean squared error (Figure 8.8b). The same procedure is followed for a problem where there are more than two variables; in which case, the problem is called a multivariant linear regression.

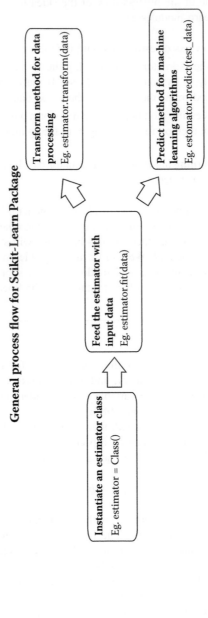

FIGURE 8.6   General Process Flow for Scikit-Learn Package.

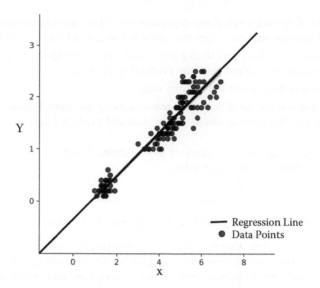

FIGURE 8.7 Linear Regression: Data Points and the Best-Fit Line.

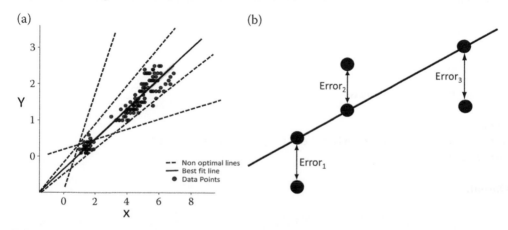

FIGURE 8.8 (a) Best Fit Line Among the Various Non-Optimal Lines and the (b) Errors Between the Data Points and the Line.

When the input has more than two variables or has higher dimensions, then the regression line is a plane for three dimensions or a hyper-plane for more than three dimensions, where the equation will be:

$$y = m_1x_1 + m_1x_1 + m_2x_2 + m_3x_3 + m_4x_4 + \cdots \cdots + m_nx_n + b$$

'In this equation, y is the dependent variable, and x_1, x_2, x_3, x_4, ..., x_n are independent variables or features. m_1, m_2, m_3, m_4, ..., m_n are the respective coefficients of the features, and these present an idea about the proportion of each feature for calculation of "y". Lastly, "b" is a bias term.

General Workflow of a Machine Learning Project

The steps to be followed in a typical Machine Learning project are given below:

1. The curation and reading of data are usually done in the Pandas dataframe.
2. In data preprocessing, if required, we execute certain visual exploratory data analysis and pre-processing of data like scaling, standardization, feature engineering, etc.

3. The assignment of the independent and dependent variables - where independent variables are the features and dependent variables - are the labels in the case of supervised learning.

4. Splitting the data into training and testing sets - where the splitting ratio is usually between 0.2 to 0.3 - depends on the user criteria and the availability of data.

5. Training a model is done using the training data.

6. Lastly, the final step is the prediction of test data using the trained model and the evaluation of the model based on the original results of test data and predicted data.

Now that we are aware of the basic overflow of any Machine Learning project, let us train our first Machine Learning algorithm or linear regression.

Implementation of Linear Regression Using Scikit-Learn

In this segment, we will use data wherein there are ten independent variables such as age, sex, BMI, blood pressure, and six blood serum measures (i.e. T-cells), low-density lipoproteins, high-density lipoproteins, thyroid-stimulating hormones, lamotrigine prescription, and blood sugar of 442 patients (Efron et al., 2004). The dependent variable is a quantitative measure of disease progression one year after baseline. The data is publicly available at https://www4.stat.ncsu.edu/~boos/var.select/diabetes.html, and is also provided with the supplementary materials of this chapter in the "tsv" text format named as "diabetes,tsv".

Loading Dataset

Let us load the dataset in Pandas.

Code:
```
dataset = pd.read_csv('diabetes.txt',delimiter='\t')
dataset.head()
```

Output:

TABLE 8.1

Disease Prognosis Using Diabetic Measurement Data

	Age	Sex	BMI	BP	S1	S2	S3	S4	S5	S6	Y
0	59	2	32.1	101.0	157	93.2	38.0	4.0	4.8598	87	151
1	48	1	21.6	87.0	183	103.2	70.0	3.0	3.8918	69	75
2	72	2	30.5	93.0	156	93.6	41.0	4.0	4.6728	85	141
3	24	1	25.3	84.0	198	131.4	40.0	5.0	4.8903	89	206
4	50	1	23.0	101.0	192	125.4	52.0	4.0	4.2905	80	135

In this example, we are presented with 11 columns, and the features are already discussed in Table 8.1. We will use the Pandas ".describe()" method to view some simple statistics about this data.

Code:
```
dataset.describe().T
```

	Count	Mean	Std	Min	25%	50%	75%	Max
AGE	442.0	48.518100	13.109028	19.0000	38.2500	50.00000	59.0000	79.000
SEX	442.0	1.468326	0.499561	1.0000	1.0000	1.00000	2.0000	2.000
BMI	442.0	26.375792	4.418122	18.0000	23.2000	25.70000	29.2750	42.200
BP	442.0	94.647014	13.831283	62.0000	84.0000	93.00000	105.0000	133.000
S1	442.0	189.140271	34.608052	97.0000	164.2500	186.00000	209.7500	301.000
S2	442.0	115.439140	30.413081	41.6000	96.0500	113.00000	134.5000	242.400
S3	442.0	49.788462	12.934202	22.0000	40.2500	48.00000	57.7500	99.000
S4	442.0	4.070249	1.290450	2.0000	3.0000	4.00000	5.0000	9.090
S5	442.0	4.641411	0.522391	3.2581	4.2767	4.62005	4.9972	6.107
S6	442.0	91.260181	11.496335	58.0000	83.2500	91.00000	98.0000	124.000
Y	442.0	152.133484	77.093005	25.0000	87.0000	140.50000	211.5000	346.000

Output:

At this point, the independent variable has a mean of 152.13 and a standard deviation of 77.09. Therefore, a trained model having a root mean square error of less than 77.09 will be above the baseline. Next, we will separate the features and the labels. Here, labels are continuous values, as it is a regression problem:

Code:

```
X=dataset.iloc[:,:-1]
y = dataset.iloc[:,-1]
```

The "*X*" variable contains all of the columns except for the last one, which is separated and stored in the "*y*" variable.

Train-Test Split

Code:

```
from sklearn.model_selection import train_test_split
X_train, X_test, y_train, y_test = train_test_split(X, y, test_size=0.2,
                                                    random_state=110)
```

In the code above, we have used a module called the "train_test_split()" from sklearn's model_selection library. The "train_test_split()" module has the features, labels, and splits ratio, where we have passed *X*, *y*, and 0.2, respectively. We have also used a variable called "random_state". This is because the module "train_test_split" assigns data points to the training set and test set randomly. Passing the "random_state" argument will keep the randomness constant, and we will acquire the same split every time with a random state value. This module will produce four variables (i.e. "X_train"), which contains a set of features that will be used for training the model with the "y_train". This, in turn, contains their respective labels, and the "X_test" and "y_test" are part of test data. The model will be evaluated using these.

Training Model

We have previously discussed the basic flow for training a model using Scikit-Learn. Below is the exccution of the steps:

Code:
```
from sklearn.linear_model import LinearRegression
regressor = LinearRegression()
regressor.fit(X_train, y_train)
```

We first import the "LinearRegression" class from sklearn's linear_model library and instantiate it as "regressor" in the second line. Then we pass our training features and respective labels to the ".fit()" method of the class. The "fit()" method will train the model using the provided data. The sklearn package is rather user-friendly that, with only three lines of code, we have successfully trained our model. After training the model, the next step is to evaluate the model using the test data.

Model Evaluation

The model classes of sklearn have a ".predict()" method which predicts values based on the features provided. We will utilize this method to forecast values based on "X_test" data and compare it with "y_test" labels.

Code:
```
predictions=regressor.predict(X_test)
plt.scatter(y_test, predictions,color='k')
plt.xlabel('Test Data')
plt.ylabel('Predicted Data')
plt.title('Test Data Vs. Predicted Data')
plt.show()
```

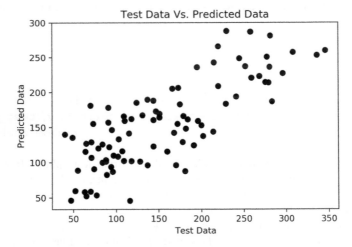

FIGURE 8.9 Comparision of Predicted Values and Original Values.

Output:
In the code above, we have plotted the predicted values and original values using Matplotlib's scatter method. For a model that has zero error, we will gain a diagonal line, whereas, in the plot generated in Figure 8.9, values are distributed around the diagonal but present a trend towards the diagonal. Visual evaluation of this model does not provide exact insights about the errors in the model. In this case, we use quantitative measures. Quantitatively, the errors are the differences between the predicted values and the original values. A standard measurement of the errors is the mean or average of all those discrepancies. In evaluating models that predict continuous values, three types of quantitative values

are used: the mean absolute error or the average sum error, the mean square error, and the root means square error. These errors can be calculated using the methods from sklearn's metrics library.

Code:
```
Form sklearn import metrics
print('Mean Absolute Error:',
        metrics.mean_absolute_error(y_test,predictions))
print('Mean Squared Error:',
        metrics.mean_squared_error(y_test, predictions))
print('Root Mean Squared Error:',
        np.sqrt(metrics.mean_squared_error(y_test, predictions)))
```

Output:
```
Mean Absolute Error: 39.22466065716574
Mean Squared Error: 2228.0908310594655
Root Mean Squared Error: 47.202657033894454
```

While discussing multivariable linear regression, we have learned in the introduction section about the equation for the hype-plane in which the "y" values can be predicted. The equations contain the intercept and the coefficients for every feature. These coefficients can provide us with an idea about the importance of various features. We can retrieve the intercept and coefficient using model objects ".intercept_" and ".coef_".

Code:
```
#To retrieve the intercept:
print(regressor.intercept_)
#For retrieving the slope:
print(regressor.coef_)
```

Output:
```
-307.57104546505445
[1.40608997e-01   -2.56128390e+01  5.08771783e+00 1.19293496e+00
-1.08819176e+00    8.03227722e-01  4.24108443e-02 4.53744471e+00
 6.51304992e+01    3.71060693e-01]
```

The code above prints the intercept and coefficient of the trained model. Let us map these with the features.

Code:
```
coeff_df = pd.DataFrame(regressor.coef_.T, X.columns, columns=['Coefficient'])
coeff_df
```

TABLE 8.2

Features with Their Coefficients

	Coefficient
Age	0.140609
Sex	−25.612839
BMI	5.087718
BP	1.192935
S1	−1.088192
S2	0.803228
S3	0.042411
S4	4.537445
S5	65.130499
S6	0.371061

Output:

The feature S5 has the height coefficient (i.e. 65.13). This can be interpreted as the prescription of the drug lamotrigine can increase the risk of the progression of the disease by around 65 times if the remaining features are unit values (Table 8.2).

TABLE 8.3

Galton Height Data

	Family	Father	Mother	Gender	Height	Kids
0	1	78.5	67.0	M	73.2	4
1	1	78.5	67.0	F	69.2	4
2	1	78.5	67.0	F	69.0	4
3	1	78.5	67.0	F	69.0	4
4	2	75.5	66.5	M	73.5	4

Predicting Child Height Based on Parents Height

In this project, we will use famous Galton height data (Galton, 2017, 1886) in which Francis Galton has recorded the heights of parents and their children and introduced the concept of regression. The data is provided in supplementary files along with the Jupyter Notebook for this chapter. Let us read the dataset below:

Code:
```
dataset = pd.read_csv('Galton.txt',delimiter='\t')
dataset.head()
```

Output:

The dataset consists of five columns. The first column contains the family id; the second and third columns are the heights of father and mother, respectively; the next two columns are gender and height of the child; the last column is the number of siblings (Table 8.3).

Code:
```
dataset.describe().T
```

Output:

The dataset has 898 observations. Next, we will execute a certain visual inspection of the data using Seaborn's pairplot method (Table 8.4).

TABLE 8.4

Brief Description of Galton Height Dataset

	Count	Mean	Std	Min	25%	50%	75%	Max
Father	898.0	69.232851	2.470256	62.0	68.0	69.0	71.0	78.5
Mother	898.0	64.084410	2.307025	58.0	63.0	64.0	65.5	70.5
Height	898.0	66.760690	3.582918	56.0	64.0	66.5	69.7	79.0
Kids	898.0	6.135857	2.685156	1.0	4.0	6.0	8.0	15.0

Code:

```
sns.pairplot(dataset,hue='Gender')
```

Output:

Figure 8.10 illustrates that the distribution of features, according to gender, is strikingly different. Therefore, we shall build two models - one for predicting boys' heights and another for predicting girls' heights. To achieve this, we will split our dataset based on gender.

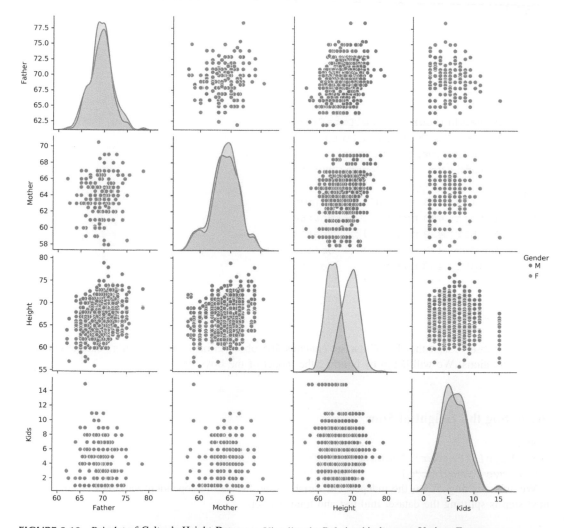

FIGURE 8.10   Pairplot of Galton's Height Dataset to Visualize the Relationship between Various Features.

Code:
```
boys = dataset[dataset['Gender']=='M']
boys.info()
```

Output:
```
<class 'pandas.core.frame.DataFrame'>
Int64Index: 465 entries, 0 to 894
Data columns (total 6 columns):
 #     Column    Non-Null Count    Dtype
---    ------    --------------    -----
 0     Family    465 non-null      object
 1     Father    465 non-null      float64
 2     Mother    465 non-null      float64
 3     Gender    465 non-null      object
 4     Height    465 non-null      float64
 5     Kids      465 non-null      int64
dtypes: float64(3), int64(1), object(2)
memory usage: 25.4+ KB
```

Code:
```
girls = dataset[dataset['Gender']=='F']
girls.info()
```

Output:
```
<class 'pandas.core.frame.DataFrame'>
Int64Index: 433 entries, 1 to 897
Data columns (total 6 columns):
 #     Column  Non-Null Count  Dtype
---    ------  --------------  -----
 0     Family  433 non-null    object
 1     Father  433 non-null    float64
 2     Mother  433 non-null    float64
 3     Gender  433 non-null    object
 4     Height  433 non-null    float64
 5     Kids    433 non-null    int64
dtypes: float64(3), int64(1), object(2)
memory usage: 23.7+ KB
```

We can observe that the "boys" dataset contains 465 observations, and the "girls" dataset contains 433 observations. We will begin by building the model of the boys first:

Predicting the Height of Sons

Code:
```
X = boys[['Father','Mother']]
y = boys[['Height']]
```
We are assigning "*X*" as the independent variables and "*y*" as the dependent variables or labels. The next step is splitting the dataset into test and train datasets.

Code:
```
from sklearn.model_selection import train_test_split
X_train, X_test, y_train, y_test = train_test_split(X, y, test_size=0.2,
random_state=110)
```

Next, we import the LinearRegression module and train the model.

Code:
```
from sklearn.linear_model import LinearRegression
regressor = LinearRegression()
regressor.fit(X_train, y_train)
```

Once the model is trained, let us retrieve the intercept and coefficient of the model.
Code:
```
print(regressor.intercept_)
coeff_df = pd.DataFrame(regressor.coef_.T, X.columns, columns=['Coefficient'])
coeff_df
```

Output:
```
[19.69844015]
```

TABLE 8.5

Coefficients for Boys Height Prediction Model

	Coefficient
Father	0.421263
Mother	0.317899

Therefore, according to the intercept and the coefficient, we can now compute an estimated equation for the calculation of a boy's height based on his parents' heights (Table 8.5):

$$\text{Boys's height} = 19.69 + (0.42*\text{Father's Height}) + (0.31*\text{Mother's height})$$

We have a prediction of the boy's height based on the equation above.

Consequently, let us evaluate the model using the three quantitative metrics:

Code:
```
from sklearn import metrics
print('Mean Absolute Error:', metrics.mean_absolute_error(y_test,
predictions))
print('Mean Squared Error:', metrics.mean_squared_error(y_test, predictions))
print('Root Mean Squared Error:', np.sqrt(metrics.mean_squared_error
(y_test, predictions)))
```

Output:
```
Mean Absolute Error: 1.7284131207951459
Mean Squared Error: 4.724556054413049
Root Mean Squared Error: 2.1736043923430612
```

We can observe that the root mean squared error is 2.17; therefore, our model has an error of ± 2.17 inches. Next, we will study how to make predictions on new data. Supposing we have the values of the heights of father and mother, and it is our goal to predict the height of their son.

Code:
```
father_height=70
mother_height=65
son_height=regressor.predict(np.array([[father_height,mother_height]]))
print(son_height)
```

Output:
```
[[69.85026337]]
```

We have to use the height of father and mother in the ".predict()" method of the model class. It should be noted that we are passing the values as nested arrays, and this is because the predict method takes an array or list of rows or observations to predict the outcomes.

Predicting the Height of Daughters

In this section, we will build the model for predicting the height of the girls. The procedure is similar as before. First, we will assign the dependent and independent values. Second, we will split the dataset into the test and train data, and then train the model using the "fit()" method and training data. Lastly, we will predict values using the test sets' independent features.

Code:
```
X = girls[['Father','Mother']]
y = girls[['Height']]
X_train, X_test, y_train, y_test = train_test_split(X, y, test_size=0.2,
random_state=110)
regressor = LinearRegression()
regressor.fit(X_train, y_train)
predictions=regressor.predict(X_test)
```

As we have accomplished in the case of sons, we will also try to compute an equation based on the intercepts and coefficient for the calculation of the height of daughters.

Code:
```
#To retrieve the intercept:
print(regressor.intercept_)
coeff_df = pd.DataFrame(regressor.coef_.T, X.columns, columns=['Coefficient'])
coeff_df
```

Output:
```
[17.53100477]
```

TABLE 8.6

Intercept and Coefficient of Features for the Daughter Height Prediction Model

	Coefficient
Father	0.407072
Mother	0.287763

Therefore, the estimated equation for a daughter's height will be (Table 8.6):

Daughter height = 17.53 + (0.40 * Father Height) + (0.28 * Mother Height)

Let us evaluate this model using the three metrics:

Code:
```
print('Mean Absolute Error:', metrics.mean_absolute_error(y_test,predictions))
print('Mean Squared Error:', metrics.mean_squared_error(y_test, predictions))
print('Root Mean Squared Error:', np.sqrt(metrics.mean_squared_error
(y_test, predictions)))
```

Output:
```
Mean Absolute Error: 1.5923427908120276
Mean Squared Error: 4.102869748180959
Root Mean Squared Error: 2.0255541829783175
```

We can identify that the root mean squared error is 2.02; therefore, our model has an error of ± 2.02 inches for the prediction of a daughter's height.

We have examined that linear regression can not only be used for the prediction of a continuous variable, but its coefficients can also give useful information about the importance or role of each feature for the calculation of the output values. We can also deduce estimated linear equations for problems and, thus, can gain clear insights into the problems. In the next four chapters, we will study Machine Learning algorithms for supervised classification. The workflow of Machine Learning projects and the flow of sklearn methods will remain almost analogous.

Exercise

1. What is the definition of Machine Learning?
2. How are classification-based models evaluated using test sets?
3. Parameter optimization of a model is one of the most important factors for improving the performance of a model. How can the parameters be optimized?
4. What type of data is inappropriate for Machine Learning tasks?
5. Explain the general steps for a typical Machine Learning implementation.
6. What is the difference between sklearn's "fit" function for a processer and an estimator?
7. Use the linear regression model to predict "BP" using "diabetes.txt" dataset.
8. Find the features which have the maximum contribution for the prediction of "BP".

REFERENCES

Bergstra, J. & Bengio, Y. (2012). Random search for hyper-parameter optimization. *The Journal of Machine Learning Research*, *13*(1), 281–305.

Efron, B., Hastie, T., Johnstone, I., & Tibshirani, R (2004). Least angle regression *Annals of Statistics*, *32*(2), 407–499. doi:10.1214/009053604000000067. Retrieved from https://projecteuclid.org/euclid.aos/1083178935.

Galton, F. (1886). Regression towards mediocrity in hereditary stature. *The Journal of the Anthropological Institute of Great Britain and Ireland*. https://doi.org/10.2307/2841583.

Galton, F. (2017). *Galton height data*. Harvard Dataverse. https://doi.org/10.7910/DVN/T0HSJ1.

Smith, J. W., Everhart, J. E., Dickson, W. C., Knowler, W. C., & Johannes, R. S. (1988). Using the ADAP learning algorithm to forecast the onset of diabetes mellitus. In *Proceedings of the Annual Symposium on Computer Applications in Medical Care*, pp. 261–265.

9

Logistic Regression

Introduction

In the previous chapter, we have learned about linear regression where we predict continuous variables, as we have done while predicting the progress of disease and the height of sons and daughters based on the height of their parents. In this chapter, we will see the problem of classification or the prediction of categorical values. In classification, we forecast the category in which a new observation will belong based on the features of the observation.

Classifying spam or ham emails, predicting whether a tumor is benign or malignant, and predicting whether a person has a disease or not are some of the examples of classification problems. These are often called binary classification - in which we have two classes. Generally, we denote two classes as "0" and "1". For example, if a tumor is benign, then it is assigned to a class "0", and if it is malignant, then it is assigned to the class "1". Therefore, the model has to predict values "0" or "1" as the output. The equation of linear regression can have output ranging from –(ve) infinity to +(ve) infinity, so we cannot use that function. Therefore, for classification, we use the logistic function called the sigmoid function. It has an S-shaped curve as shown in Figure 9.1.

As shown in Figure 9.1, for any value of x, the output will always be between "0" and "1". Supposing we want to predict if a tumor is benign (i.e. "0" or malignant and "1"), then the sigmoid function will give values ranging from 0 to 1. Therefore, we have to set a threshold to acquire only 0 and 1 values. The most common threshold is 0.5, and if a value is <0.5 then it is assigned to the 0 class, and if the value is greater than or equal to 0.5, then it classified to the 1 class. The output values of the sigmoid function are also considered as probabilities of falling in the particular class.

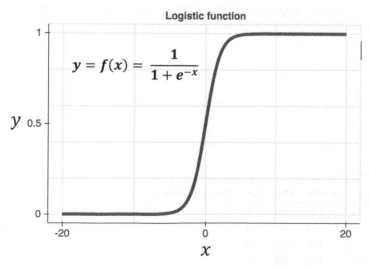

FIGURE 9.1 The Logistic or Sigmoid Function.

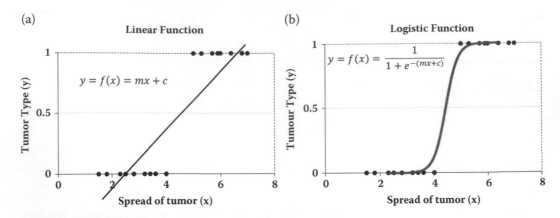

FIGURE 9.2   Linear Regression Function and the Logistic Regression Function.

The normal linear regression model is incapable of executing this classification task accurately but can be put into the sigmoid function to achieve the required goal. Let us understand this further with an example:

Supposing we have an imaginary dataset of tumor malignancy versus tumor spread. The tumor spread is an independent variable based on which our function predicts if the tumor is malignant or not. Value "1" is assigned for malignant tumors, and "0" represents benign tumors. Both plots in Figure 9.2 illustrate the relationship between the two variables. Figure 9.2a shows the fitting of the linear regression function which, as we have previously discussed, is a bad fit. Instead, we can transform our linear regression to a logistic regression curve. In Figure 9.2b, we can notice that our transformed linear regression function using the logistic function provides a value between 0 and 1. Along with the coefficients, logistic regression also presents a probability value, which shows the confidence of the model over its prediction. Next, we will implement logistic regression using the sklearn library.

Implementation of Logistic Regression Using Sklearn

The dataset we will use in this part consists of 13 independent features for the prediction of heart disease (Detrano et al., 1989). The dataset can be found in the UCI Machine Learning Repository (Dua and Graff, 2017), which is a popular place to find real-world datasets for Machine Learning tasks. The dataset is also provided as a supplementary file with this chapter. We will use this dataset as a practice dataset to examine other classification algorithms along with other examples of real-world data. Let us load this dataset to the dataframe.

Code:

```
dataset = pd.read_csv('heart.csv')
dataset.head()
```

TABLE 9.1

Heart Disease Dataset

	Age	Sex	cp	trestbps	chol	fbs	restecg	thalach	exang	oldpeak	slope	ca	thal	Target
0	63	1	3	145	233	1	0	150	0	2.3	0	0	1	1
1	37	1	2	130	250	0	1	187	0	3.5	0	0	2	1
2	41	0	1	130	204	0	0	172	0	1.4	2	0	2	1
3	56	1	1	120	236	0	1	178	0	0.8	2	0	2	1
4	57	0	0	120	354	0	1	163	1	0.6	2	0	2	1

Output:

We can observe that the dataset contains features like age, sex, chest pain (cp), resting blood pressure **(trestbps),** serum cholesterol (chol), fasting blood sugar (fbs), resting electrocardiographic results **(restecg),** maximum heart rate achieved **(thalach),** exercise-induced angina (exang), ST depression induced by exercise relative to rest (oldpeak), the slope of the peak exercise ST segment (slope), number of major vessels (0–3) colored by fluoroscopy (ca), and, lastly, the target - where "0" is normal individual and "1" is for heart disease (Table 9.1).

Code:
```
dataset['target'].value_counts()
```
Output:
```
1   165
0   138
Name: target, dtype: int64
```
The ".value_counts()" method of dataframe provides the frequency of the unique values of a series. As explained here, we are presented with 165 positive instances and 138 negative instances. Before training a classifier, we have to set a baseline accuracy. When falling below baseline accuracy, the model is not acceptable. Baseline accuracy generally is the percentage of the major class in the dataset. In this example, the baseline accuracy will be 165/303 or 54%.

Visualization of the continuous variable and its relation with the other variables and classes provides us with an overview of the dataset. We have Seaborn's pairplot to achieve this objective.

Code:
```
sns.pairplot(dataset
[['age','trestbps','chol','thalach','oldpeak','target']],

hue = 'target')
```
Output:
Figure 9.3 shows that the target classes are overlapping with one another. Dependent variables have little to no correlations among themselves. However, variations among the classes are distinguishable, which is an advantage.

The next step is to divide the data into independent and dependent variables. All of the columns, except for the last column, are independent variables and features, while the last column or the target column is the label.

Code:
```
X = dataset.iloc[:,:-1]
print(X.info())
y = dataset.iloc[:,-1]
```
Output:
```
<class 'pandas.core.frame.DataFrame'>

RangeIndex: 303 entries, 0 to 302

Data columns (total 13 columns):

#   Column   Non-Null Count       Dtype

---  ------   --------------       -------

0   age      303 non-null         int64

1   sex      303 non-null         int64
```

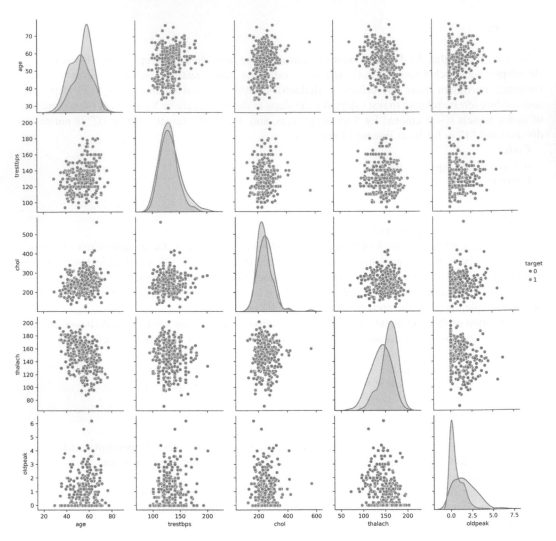

FIGURE 9.3   Pairplot for Heart Disease Dataset.

```
2    cp         303 non-null    int64

3    trestbps   303 non-null    int64

4    chol       303 non-null    int64

5    fbs        303 non-null    int64

6    restecg    303 non-null    int64

7    thalach    303 non-null    int64

8    exang      303 non-null    int64

9    oldpeak    303 non-null    float64
```

```
10   slope       303 non-null     int64

11   ca          303 non-null     int64

12   thal        303 non-null     int64

dtypes: float64(1), int64(12)

memory usage: 30.9 KB

None
```
Therefore, we have 13 features and 303 observations with probably no null values, because the "Non-Null count" column has an equal count for all features.

Train-Test Split

Code:
```
from sklearn.model_selection import train_test_split
X_train, X_test, y_train, y_test = train_test_split(X, y, test_size = 0.25,
random_state = 0)
```

Training the Logistic Regression Model

The "LogisticRegression" class is present in the "linear_model" library of the sklearn package.
   Code:
```
from sklearn.linear_model import LogisticRegression
classifier = LogisticRegression()
classifier.fit(X_train, y_train)
```
In this example, we built the classifier object and trained it on the training set using the "fit()" method of the "logistic_regression" class.

Evaluation of Model

To evaluate a model, we compare newly generated values with the original ones. In the case of a classification task, these values are compared with one another in a confusion matrix. Recall from chapter 8, a confusion matrix is constituted of the values of true positives, true negatives, false positives, and false negatives (Figure 8.3, Table 9.2).
   Code:
```
y_pred = classifier.predict(X_test)
from sklearn.metrics import confusion_matrix, accuracy_score
cm = confusion_matrix(y_test, y_pred)
print(cm)
accuracy_score(y_test, y_pred)
```

TABLE 9.2

Confusion Matrix

	Actual Positive (1)	**Actual Negative (0)**
Predicted Positive (1)	True Positives (TP)	False positives (FP)
Predicted Negative (0)	False Negatives (FN)	True Negatives (TN)

Output:
```
[[24  9]
 [ 3 40]]
0.8421052631578947
```
In the set of codes above, we first predict outcomes using the test set. The sklearn package provides various methods for the evaluation of the models. As we have learned in the previous chapter, the metrics library holds all of these methods. These methods use the predicted values and the original values and subsequently return the evaluation outcome. The "confusion_matrix" and "accuracy_score" are two of these methods. Here, the accuracy of our model is 0.84 or 84%. In chapter 8, we have also discussed various mathematical models for the evaluation of a model like Precision, Recall, F1 Score, etc.

The metrics library has a "classification_report" method that calculates all of the above-stated values and provides it in a convenient and presentable way.

Code:
```
from sklearn.metrics import classification_report
print(classification_report(y_test, y_pred))
```
Output:

	precision	recall	f1-score	support
0	0.89	0.73	0.80	33
1	0.82	0.93	0.87	43
accuracy			0.84	76
macro avg	0.85	0.83	0.83	76
weighted avg	0.85	0.84	0.84	76

This method also uses the predicted and the original values to make the report. This report contains respective values in each class, as well as the weighted average, or the values for the overall model.

Retrieving Intercept and Coefficient

Like linear regression, logistic regression also provides the intercept and the coefficients.

Code:
```
#To retrieve the intercept:
print(classifier.intercept_)
#For retrieving the slope:
print(classifier.coef_)
```
Output:
```
[0.01939654]
[[ 0.0122945 -1.54987762  0.81951552 -0.00694753 -0.00419397 -
0.39621953
   0.27023142  0.02894972 -0.78611352 -0.56529797  0.25414543 -
0.7385799
  -0.7680082 ]]
```
Code:
```
coeff_df = pd.DataFrame(classifier.coef_.T, X.columns,

columns=['Coefficient'])
coeff_df =coeff_df.sort_values(by = ['Coefficient'],ascending=False)
coeff_df
```
Output:
In this part, we have sorted the features according to their coefficients, and it appears that chest pain is one of the major symptoms for heart diseases, followed by the resting electrocardiographic results (Table 9.3). Let us visualize the distribution of these features in both of the classes.

Code:
```
fig, axes = plt.subplots(nrows=1, ncols=2,figsize=(10,4))
```

TABLE 9.3

Coefficient of Features Retrieved Form Logistic Regression

	Coefficient
cp	0.819516
restecg	0.270231
slope	0.254145
thalach	0.028950
age	0.012294
chol	−0.004194
trestbps	−0.006948
fbs	−0.396220
oldpeak	−0.565298
ca	−0.738580
thal	−0.768008
exang	−0.786114
sex	−1.549878

```
sns.countplot(x="cp", hue="target", data=dataset,ax=axes[0])
sns.boxplot(y='thalach',x='target',
          data=dataset,palette="coolwarm",ax=axes[1])
fig.tight_layout()
```
Output:

Since chest pain is a categorical feature, we have to draw a count plot and indicate the count of each class in the categories where chest pain is plotted. From the count plot in Figure 9.4a, we can observe that for "0" chest pain, we have less counts for heart disease. As the category of chest pain changes to "1", the count for heart disease also increases. In Figure 9.4b, we are presented with a boxplot for the distribution of the resting electrocardiographic results versus the target. The difference in distribution is visible. Low values of resting electrocardiographic measurement indicates lesser probability of heart disease and vice versa.

Data Scaling

Data scaling causes the range of variables to become uniform. In this example, we have 12 features. Let us study their spread using Pandas' ".describe()" function.

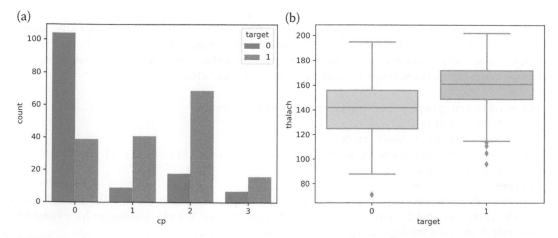

FIGURE 9.4 Comparison of the Two Most Important Features for Determining Heart Disease.

TABLE 9.4

Standard Statistics for the Heart Disease Dataset

	Count	Mean	Std	Min	25%	50%	75%	Max
age	303.0	54.366337	9.082101	29.0	47.5	55.0	61.0	77.0
sex	303.0	0.683168	0.466011	0.0	0.0	1.0	1.0	1.0
cp	303.0	0.966997	1.032052	0.0	0.0	1.0	2.0	3.0
trestbps	303.0	131.623762	17.538143	94.0	120.0	130.0	140.0	200.0
chol	303.0	246.264026	51.830751	126.0	211.0	240.0	274.5	564.0
fbs	303.0	0.148515	0.356198	0.0	0.0	0.0	0.0	1.0
restecg	303.0	0.528053	0.525860	0.0	0.0	1.0	1.0	2.0
thalach	303.0	149.646865	22.905161	71.0	133.5	153.0	166.0	202.0
exang	303.0	0.326733	0.469794	0.0	0.0	0.0	1.0	1.0
oldpeak	303.0	1.039604	1.161075	0.0	0.0	0.8	1.6	6.2
slope	303.0	1.399340	0.616226	0.0	1.0	1.0	2.0	2.0
ca	303.0	0.729373	1.022606	0.0	0.0	0.0	1.0	4.0
thal	303.0	2.313531	0.612277	0.0	2.0	2.0	3.0	3.0
target	303.0	0.544554	0.498835	0.0	0.0	1.0	1.0	1.0

Code:

```
dataset.describe().T
```

Output:

Check the mean, std, min, and max columns of the continuous features like age, trestbps, chol, thalach, and oldpeak (Table 9.4). All of these values have considerable differences in their ranges, and this can affect the training of models. To produce the data in similar ranges, we execute scaling. Sklearn has various such classes in its "preprocessing" library. Kindly recall from chapter 8 the flow of preprocessing classes of the sklearn package. First, we will observe the "StandardScaler" class, which subtracts all of the observations with their means and then divides them with standard deviation as we learned to do before applying PCA in chapter 6

Code:

```
from sklearn.preprocessing import StandardScaler
sc = StandardScaler()
X_norm = sc.fit_transform(X)
X_scaled = pd.DataFrame(X_norm,columns=dataset.iloc[:,0:-1].columns)
X_scaled.describe().T
```

Output:

Comparing Tables 9.4 and 9.5, we can see that the mean, std, min, and max values for continuous values have changed. Now, the observations are scaled and transformed into uniform distribution, which is preferred by most Machine Learning algorithms. We can produce the data for every feature into an identical range (i.e. between 0 and 1) using the "MinMaxScaler" class.

Code:

```
from sklearn.preprocessing import MinMaxScaler
sc = MinMaxScaler()
X_minmax = sc.fit_transform(X)
X_minmax = pd.DataFrame(X_minmax,columns=dataset.iloc[:,0:-1].columns)
X_minmax.describe().T
```

Output:

In Table 9.6, we can observe the values are in identical range if we refer to the columns min and max. The min value is "0", and the max value is "1" for all of the features. We will continue with the standard scalar output and train the logistic regression model on scaled data.

TABLE 9.5

Standard Scaled Value's Statistics for Heart Disease Dataset

	Count	Mean	Std	Min	25%	50%	75%	Max
age	303.0	4.690051e−17	1.001654	−2.797624	−0.757280	0.069886	0.731619	2.496240
sex	303.0	−1.407015e−16	1.001654	−1.468418	−1.468418	0.681005	0.681005	0.681005
cp	303.0	2.345026e−17	1.001654	−0.938515	−0.938515	0.032031	1.002577	1.973123
trestbps	303.0	−7.035077e−16	1.001654	−2.148802	−0.663867	−0.092738	0.478391	3.905165
chol	303.0	−1.113887e−16	1.001654	−2.324160	−0.681494	−0.121055	0.545674	6.140401
fbs	303.0	−2.345026e−17	1.001654	−0.417635	−0.417635	−0.417635	−0.417635	2.394438
restecg	303.0	1.465641e−16	1.001654	−1.005832	−1.005832	0.898962	0.898962	2.803756
thalach	303.0	−6.800574e−16	1.001654	−3.439267	−0.706111	0.146634	0.715131	2.289429
exang	303.0	−4.690051e−17	1.001654	−0.696631	−0.696631	−0.696631	1.435481	1.435481
oldpeak	303.0	2.345026e−17	1.001654	−0.896862	−0.896862	−0.206705	0.483451	4.451851
slope	303.0	−1.407015e−16	1.001654	−2.274579	−0.649113	−0.649113	0.976352	0.976352
ca	303.0	−2.345026e−17	1.001654	−0.714429	−0.714429	−0.714429	0.265082	3.203615
thal	303.0	−1.641518e−16	1.001654	−3.784824	−0.512922	−0.512922	1.123029	1.123029

TABLE 9.6

Statistics for MinMaxscaler Values of Heart Disease Dataset

	Count	Mean	Std	Min	25%	50%	75%	Max
age	303.0	0.528465	0.189210	0.0	0.385417	0.541667	0.666667	1.0
sex	303.0	0.683168	0.466011	0.0	0.000000	1.000000	1.000000	1.0
cp	303.0	0.322332	0.344017	0.0	0.000000	0.333333	0.666667	1.0
trestbps	303.0	0.354941	0.165454	0.0	0.245283	0.339623	0.433962	1.0
chol	303.0	0.274575	0.118335	0.0	0.194064	0.260274	0.339041	1.0
fbs	303.0	0.148515	0.356198	0.0	0.000000	0.000000	0.000000	1.0
restecg	303.0	0.264026	0.262930	0.0	0.000000	0.500000	0.500000	1.0
thalach	303.0	0.600358	0.174849	0.0	0.477099	0.625954	0.725191	1.0
exang	303.0	0.326733	0.469794	0.0	0.000000	0.000000	1.000000	1.0
oldpeak	303.0	0.167678	0.187270	0.0	0.000000	0.129032	0.258065	1.0
slope	303.0	0.699670	0.308113	0.0	0.500000	0.500000	1.000000	1.0
ca	303.0	0.182343	0.255652	0.0	0.000000	0.000000	0.250000	1.0
thal	303.0	0.771177	0.204092	0.0	0.666667	0.666667	1.000000	1.0

Code:

```
from sklearn.model_selection import train_test_split
X_train, X_test, y_train, y_test = train_test_split(X_norm, y, test_size =

0.2, random_state = 0)
from sklearn.linear_model import LogisticRegression
classifier = LogisticRegression()
classifier.fit(X_train, y_train)
y_pred = classifier.predict(X_test)
```

Once the model is trained on scaled data, we can evaluate it once again to confirm if we have improved the model:

Code:

```
from sklearn.metrics import confusion_matrix, classification_report
```

```
cm = confusion_matrix(y_test, y_pred)
print(cm)
print(classification_report(y_test, y_pred))
```
Output:
```
[[21  6]
 [3 31]]
```

	precision	recall	f1-score	support
0	0.88	0.78	0.82	27
1	0.84	0.91	0.87	34
accuracy			0.85	61
macro avg	0.86	0.84	0.85	61
weighted avg	0.85	0.85	0.85	61

By scaling the data, we gained an improvement of 1%, with respect to unscaled data. This shows that logistic regression is not too sensitive towards data scaling, unlike certain algorithms such as artificial neural networks, which are highly sensitive with data scaling. We will discuss these more in chapter 14.

Predicting a New Result

We can predict results using the "predict" method of the algorithm class. To show an example of prediction, we will study an observation from the dataset.

Code:
```
dataset.iloc[1,:]
```
Output:
```
age        37.0
sex        1.0
cp         2.0
trestbps   130.0
chol       250.0
fbs        0.0
restecg    1.0
thalach    187.0
exang      0.0
oldpeak    3.5
slope      0.0
ca         0.0
thal       2.0
target     1.0
Name: 1, dtype: float64
```
We recognize that data should be provided in a nested array, because the "predict" method requires an array of instances as an input.

Code:
```
new_data=dataset.iloc[1,:-1].values
classifier.predict(sc.transform([new_data]))
```
Output:
```
array([1], dtype=int64)
```
The model predicted that the observation belongs to class "1" (i.e. having heart disease). Logistic regression can also provide the probability of the observation for belonging to the predicted class. This can be retrieved using the "predict_proba()" method.

Code:
```
classifier.predict_proba(sc.transform([new_data]))
```
Output:
```
array([[0.27675793, 0.72324207]])
```

The output illustrates that the observation has a 27% chance of being in class "0" and a 73% chance of being in class "1". The probability of falling in class 1 is higher than 50%; therefore, the observation has been assigned to class "1".

Breast Cancer Prediction Using Logistic Regression

In this project, we will use the Breast Cancer Wisconsin (Diagnostic) Dataset (Dua and Graff, 2017; Mangasarian et al., 1995), which is publicly available in the UCI Machine Learning repository. This dataset contains 30 derived features from cell images of breast benign and malignant tumors. There are 569 observations, we will apply logistic regression to classify these observations as a benign tumor or a malignant tumor.

Code:

```
dataset = pd.read_csv('Breast_Cancer.csv')
dataset.head().T
```

Output:

	0 842302	1 842517	2 84300903	3 84348301	4 84358402
id					
diagnosis	M	M	M	M	M
radius_mean	17.99	20.57	19.69	11.42	20.29
texture_mean	10.38	17.77	21.25	20.38	14.34
perimeter_mean	122.8	132.9	130	77.58	135.1
area_mean	1001	1326	1203	386.1	1297
smoothness_mean	0.1184	0.08474	0.1096	0.1425	0.1003
compactness_mean	0.2776	0.07864	0.1599	0.2839	0.1328
concavity_mean	0.3001	0.0869	0.1974	0.2414	0.198
concave points_mean	0.1471	0.07017	0.1279	0.1052	0.1043
symmetry_mean	0.2419	0.1812	0.2069	0.2597	0.1809
fractal_dimension_mean	0.07871	0.05667	0.05999	0.09744	0.05883
radius_se	1.095	0.5435	0.7456	0.4956	0.7572
texture_se	0.9053	0.7339	0.7869	1.156	0.7813
perimeter_se	8.589	3.398	4.585	3.445	5.438
area_se	153.4	74.08	94.03	27.23	94.44
smoothness_se	0.006399	0.005225	0.00615	0.00911	0.01149
compactness_se	0.04904	0.01308	0.04006	0.07458	0.02461
concavity_se	0.05373	0.0186	0.03832	0.05661	0.05688
concave points_se	0.01587	0.0134	0.02058	0.01867	0.01885
symmetry_se	0.03003	0.01389	0.0225	0.05963	0.01756
fractal_dimension_se	0.006193	0.003532	0.004571	0.009208	0.005115
radius_worst	25.38	24.99	23.57	14.91	22.54
texture_worst	17.33	23.41	25.53	26.5	16.67
perimeter_worst	184.6	158.8	152.5	98.87	152.2
area_worst	2019	1956	1709	567.7	1575
smoothness_worst	0.1622	0.1238	0.1444	0.2098	0.1374
compactness_worst	0.6656	0.1866	0.4245	0.8663	0.205
concavity_worst	0.7119	0.2416	0.4504	0.6869	0.4
concave points_worst	0.2654	0.186	0.243	0.2575	0.1625
symmetry_worst	0.4601	0.275	0.3613	0.6638	0.2364
fractal_dimension_worst	0.1189	0.08902	0.08758	0.173	0.07678
Unnamed: 32	NaN	NaN	NaN	NaN	NaN

The dataset is constituted of 33 features. We can observe that the features named "id" and "Unnamed:32" are unwanted. Let us eliminate these.

Code:
```
dataset = dataset.drop(['id','Unnamed: 32'],axis=1)
dataset['diagnosis'].value_counts()
```
Output:
```
B    357
M    212
Name: diagnosis, dtype: int64
```
We have acquired 357 instances of benign as denoted by "B" and 212 instances of malignant as denoted by "M". Therefore, the base accuracy of the model will be 62.7%. In order to improve the accuracy of the model, we will follow the general workflow of a Machine Learning project.

Code:
```
X = dataset.iloc[:,1:]
y = dataset.iloc[:,0]
```
We separated the data into independent variables and dependent variables.

Code:
```
from sklearn.preprocessing import StandardScaler
sc = StandardScaler()
X_norm = sc.fit_transform(X)
```
Next, we applied StandardScaler to the features. Moreover, the data will be split into a training and a test set, and then the model will be trained using the training set.

Code:
```
from sklearn.model_selection import train_test_split
X_train, X_test, y_train, y_test = train_test_split(X_norm, y, test_size =
0.2, random_state = 0)
from sklearn.linear_model import LogisticRegression
classifier = LogisticRegression(random_state = 0)
classifier.fit(X_train, y_train)
```

Model Evaluation

Code:
```
y_pred = classifier.predict(X_test)
from sklearn.metrics import confusion_matrix, classification_report
cm = confusion_matrix(y_test, y_pred)
print(cm)
print(classification_report(y_test, y_pred))
```
Output:
```
[[65  2]

 [ 2 45]]
```

	precision	recall	f1-score	support
B	0.97	0.97	0.97	67
M	0.96	0.96	0.96	47
accuracy			0.96	114
macro avg	0.96	0.96	0.96	114
weighted avg	0.96	0.96	0.96	114

In this classification task, we have achieved an accuracy rate of 96%, which is much better than the base accuracy which used to be 62.7%. Now, we will identify the informative features with a higher impact on the prediction of tumor category by retrieving the coefficient.

Code:
```
coeff_df = pd.DataFrame(classifier.coef_.T, X.columns,
columns=['Coefficient'])
coeff_df =coeff_df.sort_values(by=['Coefficient'],ascending=False)
coeff_df
```
Output:

From the coefficient table, it can be detected that the features "radius_se" and "texture_worst" have a major contribution to determining the malignancy of breast tumors (Table 9.7). Radius determines the size of the cell, whereas the texture is the contrast of the cell images. For more information on features, readers can refer to https://archive.ics.uci.edu/ml/datasets/Breast+Cancer+Wisconsin+(Diagnostic).

Code:
```
fig, axes = plt.subplots(nrows=1, ncols=2,figsize=(10,4))
sns.boxplot(x="radius_se", y="diagnosis", data=dataset,ax=axes[0])
sns.boxplot(x='texture_worst',y='diagnosis',
            data=dataset,ax=axes[1])
fig.tight_layout()
```

TABLE 9.7

Coefficient of Features for Breast Cancer Prediction

	Coefficient
radius_se	1.376258
texture_worst	1.058625
area_se	0.994695
concave points_worst	0.987129
radius_worst	0.922528
area_worst	0.909691
concavity_worst	0.870952
concave points_mean	0.843549
perimeter_se	0.812926
perimeter_worst	0.758439
concavity_mean	0.676390
fractal_dimension_worst	0.597940
smoothness_worst	0.536471
symmetry_worst	0.522836
texture_mean	0.494943
area_mean	0.395679
symmetry_mean	0.334403
radius_mean	0.325119
perimeter_mean	0.317433
concave points_se	0.225080
smoothness_mean	0.188723
compactness_worst	0.028397
texture_se	−0.045093
concavity_se	−0.102279
symmetry_se	−0.141223
fractal_dimension_mean	−0.209551
smoothness_se	−0.257041
compactness_mean	−0.436253
compactness_se	−0.671800
fractal_dimension_se	−0.872027

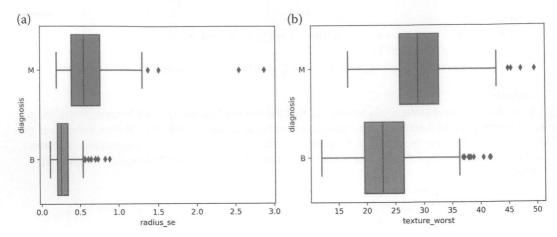

FIGURE 9.5   Data Distribution for Features "radius_se" and "texture_worst" versus Target Classes.

Output:
In Figure 9.5, we can visualize the spread of the feature "radius_se" and "texture_worst" for the malignant and benign categories. Malignant cells are, on average, larger than benign cells.

In this chapter, we have discussed logistic regression and its implementation using sklearn. We used various mathematical models to evaluate the classification models. Logistic regression is widely used for determining the important independent features in a study. Logistic regression falls in the category of linear function classifiers, and it is difficult to be trained on data where classes are not linearly separable. This is one of the limitations of logistic regression. In the next chapter, we will discuss one of the simplest types of classification algorithms – the *k*-nearest neighbor algorithm.

Exercise

1. Why can linear regression not be used for classification tasks?
2. How can a sigmoid function classify labels?
3. Why is data scaling important in Machine Learning?
4. Perform an exploratory data analysis on the breast cancer dataset using various plots.
5. Load the penguin dataset from the Seaborn pre-installed datasets and use logistic regression to classify the penguins to their species.
6. Determine the feature that contributes the most in classifying the penguins.

REFERENCES

Detrano, R., Janosi, A., Steinbrunn, W., Pfisterer, M., Jakob Schmid, J., Sandhu, S., Guppy, K. H., Lee, S., & Froelicher, V. (1989). International application of a new probability algorithm for the diagnosis of coronary artery disease. *The American Journal of Cardiology.* https://doi.org/10.1016/0002-9149(89) 90524-9

Dua, D., & Graff, C. (2017). *{UCI} Machine learning repository.* http://archive.ics.uci.edu/ml

Mangasarian, O. L., Street, W. N., & Wolberg W. H. (1995). Breast cancer diagnosis and prognosis via linear programming. *Operations Research.* https://doi.org/10.1287/opre.43.4.570

10

K-Nearest Neighbors (K-NN)

Introduction

K-nearest neighbor, or *K*-NN, is a simple supervised learning algorithm. For a new instance, the *K*-NN finds the nearest data points or the "nearest neighbors" from the dataset through which the model is trained. For a *K*-NN, the algorithm considers the "*k*" number of the nearest neighbors for the new entry. "*K*" is generally defined by users. The prediction for the new data point is simply the class of its "*k*" nearest data points.

"A man is known by the company he keeps" - this proverb is the simplest analogy for the *K*-NN algorithm. To illustrate, while diagnosing a new case, a doctor tries to find cases with similar symptoms from the past. Therefore, symptoms can be considered as features, and the labels are diagnosis or treatments for this problem. In *K*-NN, the model representation is the sorted training dataset, and no special learning is required. As demonstrated in Figure 10.1, we have data points for two groups denoted by the blue and yellow colors. If we choose *K* = 3 and create a new instance for forecast, then the model will check the three nearest data points and assign the new data point to the majority of the nearest class. From the figure, we can see that choosing the *k*-value is an important factor in deciding the behavior of the model. Low values of *K* may compromise the accuracy of the model, since noise can have a more significant effect on the model, while greater values of *k* cause the *K*-NN algorithm to be computationally intensive. Therefore, a lower value of *k* that generates a higher accuracy rate is chosen. Selecting an optimal value of *k* is very crucial, and we will discuss this in more detail in the project section.

Different types of distance measures are used to find which are the nearest neighbors. The most common distance measure is in the Euclidean distance. Figure 10.2 shows the approach for the calculation of Euclidean distance, and it can be carried forward to higher dimensions.

Hamming distance is the sum of the absolute difference between unit vectors.

In *K*-NN, the dimensionality problem arises when the total number of features increases. The problem in high-dimension space is that all of the points will be distant from one another and their neighbors.

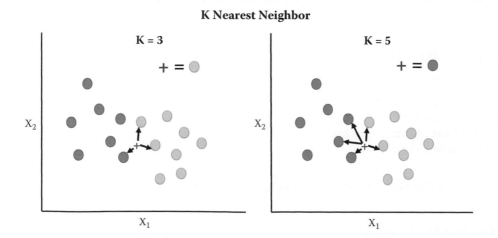

FIGURE 10.1   *K*-Nearest Neighbor Algorithm.

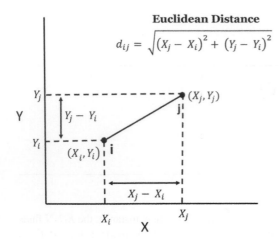

FIGURE 10.2 Calculation of Euclidean Distance.

Some of the advantages of *K*-NN include the following:

1. The algorithm is simple and easy to implement.
2. There is no need of to build a model.
3. The *K*-NN is flexible, and it can be used for classification as well as regression. In regression, it returns the mean value of *k* nearest data points.
4. *K*-NN is more effective for large training data.

Some of the limitations of K-NN are listed below:

1. When the number of *k* increases, the algorithm requires significantly high computational resources.
2. *K*-NN has a high computation cost for the prediction of new data.
3. *K*-NN is very poor in handling categorical features.

Implemention of *K*-NN Using Sklearn

First, we will apply the *K*-NN to our dummy dataset - the heart disease data. This will be easy to use with the Jupyter Notebook file which is given in the supplementary file, as most Machine Learning projects follow the same workflow.

Loading the Dataset

Code:
```
dataset = pd.read_csv('heart.csv')
dataset.head()
X = dataset.iloc[:,:-1]
y = dataset.iloc[:,-1]
```

Splitting the Dataset into the Training Set and the Test Set

Code:
```
from sklearn.model_selection import train_test_split
X_train, X_test, y_train, y_test = train_test_split(X, y, test_size = 0.25,
random_state = 0)
```

Training the *K*-NN Model on the Training Set

Sklearn's "neighbors" library contains the "KNeighborsClassifier" class. It uses "n_neighbors" parameter while initializing it. This parameter is the *k*-value. We can also pass distance measures, but here we will use the default Euclidean distance. At the first stage, we will utilize the *k*-value of "1" and determine the accuracy of the model, and then we will investigate how the optimal *k*-value could be predicted.

Code:
```
from sklearn.neighbors import KNeighborsClassifier
classifier = KNeighborsClassifier(n_neighbors = 1)
classifier.fit(X_train, y_train)
```

Evaluation with K 1

Code:
```
y_pred = classifier.predict(X_test)
from sklearn.metrics import confusion_matrix, classification_report
print('\n')
print(confusion_matrix(y_test,y_pred))
print('\n')
print(classification_report(y_test,y_pred))
```

Output:
```
[[12 21]
 [15 28]]
```

	precision	recall	f1-score	support
0	0.44	0.36	0.40	33
1	0.57	0.65	0.61	43
accuracy			0.53	76
macro avg	0.51	0.51	0.50	76
weighted avg	0.52	0.53	0.52	76

With *k* equal to "1", we attained an accuracy of 53%, which is below our baseline accuracy.

Choosing a *K*-Value

When choosing an optimal *k*-value, we will initiate the model with various *k*-values and keep track of the errors. In this case, the definition of error is the number of mispredicted or wrongly predicted labels from the test data. Next, we will plot the error data versus *k*-values and choose the *k*-values with minimal error.

Code:
```
errors = []
for i in range(1,30):
    knn = KNeighborsClassifier(n_neighbors=i)
    knn.fit(X_train,y_train)
    pred_i = knn.predict(X_test)
    errors.append(np.mean(pred_i = y_test))
plt.figure(figsize=(10,6))
plt.plot(range(1,30),error_rate,color='black')
plt.title('Error Rate vs. K Value')
plt.xlabel('K Value')
plt.ylabel('Error Rate')
```

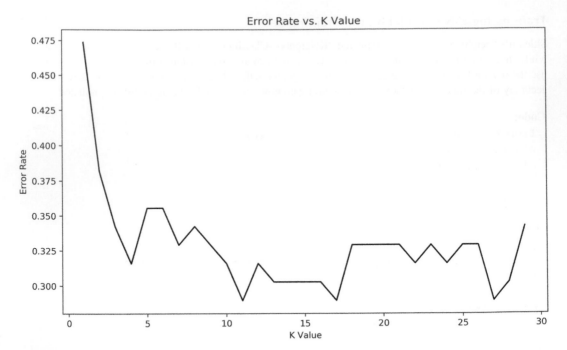

FIGURE 10.3   Error Rate versus *K*-Value Plot.

Output:
The plot of error rate versus *k*-value is often called an elbow plot because initially, small *k*-values commonly have higher error rates. The error rate gradually decreases with an increase of *k*-value, and eventually, the curve flattens, and it resembles an elbow. Figure 10.3 shows that the error is low at certain *k*-values. We will select the least *k*-value among the *k*-values producing low error. Therefore from the plot, we select the *k*-value as "11". Let us evaluate a model using the *k*-value as "11".

Code:
```
from sklearn.neighbors import KNeighborsClassifier
classifier = KNeighborsClassifier(n_neighbors = 11)
classifier.fit(X_train, y_train)
y_pred = classifier.predict(X_test)
from sklearn.metrics import confusion_matrix, classification_report
print(confusion_matrix(y_test,y_pred))
print('\n')
print(classification_report(y_test,y_pred))
```

Output:
```
[[20 13]
 [ 9 34]]
```

	precision	recall	f1-score	support
0	0.69	0.61	0.65	33
1	0.72	0.79	0.76	43
accuracy			0.71	76
macro avg	0.71	0.70	0.70	76
weighted avg	0.71	0.71	0.71	76

With the *k*-value assigned as 11, we attain an accuracy of 71%. Now, we will try the steps above with scaled data points to see if the accuracy improves.

Data Scaling

Here, the StandardScaler is used to scale the data.

Code:
```
from sklearn.preprocessing import StandardScaler
sc = StandardScaler()
X_norm = sc.fit_transform(X)
from sklearn.model_selection import train_test_split
X_train, X_test, y_train, y_test = train_test_split(X_norm, y, test_size =
0.25, random_state = 0)
```

Code:
```
errors = []
for i in range(1,15):
    knn = KNeighborsClassifier(n_neighbors=i)
    knn.fit(X_train,y_train)
    pred_i = knn.predict(X_test)
    errors.append(np.mean(pred_i != y_test))
plt.figure(figsize=(10,6))
plt.plot(range(1,15),error_rate,color='black')
plt.title('Error Rate vs. K Value')
plt.xlabel('K value')
plt.ylabel('Error Rate')
```

Output:

In Figure 10.4, we can observe that normalization has significantly changed the elbow plot. From this plot, we will use "8" as the k-value and evaluate the model for confirming accuracy.

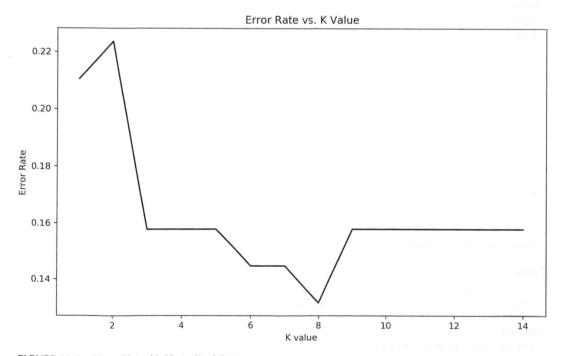

FIGURE 10.4 Elbow Plot with Normalized Dataset.

Code:
```
from sklearn.neighbors import KNeighborsClassifier
classifier = KNeighborsClassifier(n_neighbors = 8)
classifier.fit(X_train, y_train)
y_pred = classifier.predict(X_test)
from sklearn.metrics import confusion_matrix, classification_report
print(confusion_matrix(y_test,y_pred))
print('\n')
print(classification_report(y_test,y_pred))
```

Output:
```
[[27  6]
 [ 4 39]]
```

	precision	recall	f1-score	support
0	0.87	0.82	0.84	33
1	0.87	0.91	0.89	43
accuracy			0.87	76
macro avg	0.87	0.86	0.87	76
weighted avg	0.87	0.87	0.87	76

To conlcude, we obtained an accuracy of 87% by training the *K*-NN with a scaled dataset.

Predicting New Values

For predicting a new instance, we will use a data point from the dataset and pass it to the "classifier.predict()" method.

Code:
```
dataset.iloc[2,:]
```
Output:
```
age          41.0
sex           0.0
cp            1.0
trestbps    130.0
chol        204.0
fbs           0.0
restecg       0.0
thalach     172.0
exang         0.0
oldpeak       1.4
slope         2.0
ca            0.0
thal          2.0
target        1.0
Name: 2, dtype: float64
```

Code:
```
new_data=dataset.iloc[2,:-1].values
classifier.predict(sc.transform([new_data]))
```

Output:
```
array([1], dtype=int64)
```

TABLE 10.1

Indian Liver Patient Dataset

	0	1	2	3	4
Age	65	62	62	58	72
Gender	Female	Male	Male	Male	Male
Total_Bilirubin	0.7	10.9	7.3	1	3.9
Direct_Bilirubin	0.1	5.5	4.1	0.4	2
Alkaline_Phosphotase	187	699	490	182	195
Alamine_Aminotransferase	16	64	60	14	27
Aspartate_Aminotransferase	18	100	68	20	59
Total_Protiens	6.8	7.5	7	6.8	7.3
Albumin	3.3	3.2	3.3	3.4	2.4
Albumin_and_Globulin_Ratio	0.9	0.74	0.89	1	0.4
Dataset	1	1	1	1	1

Diagnosing the Liver Disease Using *K*-NN

In this study, we will use the Indian liver dataset (Venkata Ramana et al., 2011) which is publicly available in UCI Machine Learning repository (Dua & Graff, 2017). The dataset is comprised of ten independent features and has 583 observations.

Code:

```
dataset = pd.read_csv('indian_liver_patient.csv')
dataset.head().T
```

Output:

The column name of the dependent variable (i.e. the label) is currently labeled as "Dataset". We will rename it into "Target" to avoid any confusion. The "Gender" row has gender in string datatype, as algorithms can handle only numeric values, so we will convert it into categorical values - so that "Female" is annotated as "0" and "Male" is annotated as "1". Next, we will check for any missing values in the observations (Table 10.1).

Code:

```
dataset = dataset.rename({'Dataset':'Target'},axis=1)
dataset['Gender'] = dataset['Gender'].map({'Female':0,'Male':1})
dataset.isnull().sum() ## Finsing missing values
```

Output:

```
Age                          0
Gender                       0
Total_Bilirubin              0
Direct_Bilirubin             0
Alkaline_Phosphotase         0
Alamine_Aminotransferase     0
Aspartate_Aminotransferase   0
Total_Protiens               0
Albumin                      0
Albumin_and_Globulin_Ratio   4
Target                       0
dtype: int64
```

While analyzing for missing values, we obtained four "null" values in the "Albumin_and_Globulin_Ratio" feature.

Missing Value Imputation

In Machine Learning projects, it is very common to encounter missing values of features in observations. There are some useful techniques applied in dealing with the missing values; some of them are:

1. A simple technique to deal with null values is to merely drop them or remove them from the dataset. However, in the case of small datasets, this can lead to information loss.
2. Another technique is to replace the null value with the most common observation of those features, such as with the mean, median, or mode. This approach is called missing value imputation. This approach can save us from information loss, but the values will become too general. To deal with this problem, we can fill the null values based on their nearest data points or predict those values based on other independent features. Although any Machine Learning algorithm can be used for this task, *K*-NN is generally favored over others, and the process is known as the nearest neighbor imputation.

In the next two blocks of codes, we will discuss two imputation methods. First is a simple imputation, where the mean of the feature will replace missing values.

Code:
```
from sklearn.impute import SimpleImputer
imputer = SimpleImputer(strategy = 'mean')
non_nan_data = imputer.fit_transform(dataset)
dataset = pd.DataFrame(non_nan_data,columns = dataset.columns)
dataset.isnull().sum()
```

Output:
```
Age                             0
Gender                          0
Total_Bilirubin                 0
Direct_Bilirubin                0
Alkaline_Phosphotase            0
Alamine_Aminotransferase        0
Aspartate_Aminotransferase      0
Total_Protiens                  0
Albumin                         0
Albumin_and_Globulin_Ratio      0
Target                          0
dtype: int64
```
All the mutation methods are present in the "impute" library of the sklearn package. The class "SimpleImputer" is initiated as "imputer". The missing values in "Albumin_and_Globulin_Ratio" are imputed with the mean of this column. In the next block of code, we will utilize the nearest neighbor imputation.

Code:
```
from sklearn.impute import KNNImputer
imputer = KNNImputer(n_neighbors=5)
non_nan_data = imputer.fit_transform(dataset)
dataset = pd.DataFrame(non_nan_data,columns = dataset.columns)
dataset.isnull().sum()
```

Output:

```
Age                              0
Gender                           0
Total_Bilirubin                  0
Direct_Bilirubin                 0
Alkaline_Phosphotase             0
Alamine_Aminotransferase         0
Aspartate_Aminotransferase       0
Total_Protiens                   0
Albumin                          0
Albumin_and_Globulin_Ratio       0
Target                           0
dtype: int64
```

The impute library employs "KNNImputer" for performing the nearest neighbor imputation. As observed here, the null value will be replaced by the mean values of its five nearest neighbors. *K*-NN has a well-recognized application for imputing missing data.

Code:
```
dataset['Target'].value_counts()
```
Output:
```
1    416
2    167
Name: Target, dtype: int64
```
The dataset contains 416 observations for liver patients and 167 observations for non-liver patients. The dataset is highly unbalanced and has a baseline accuracy of 71%. Let us now study how the *K*-NN algorithm performs on this dataset.

Code:
```
X = dataset.iloc[:,:-1]
y = dataset.iloc[:,-1]
```

Data Scaling

Code:
```
from sklearn.preprocessing import StandardScaler
sc = StandardScaler()
X_norm = sc.fit_transform(X)
```

Splitting the Dataset into the Training Set and the Test Set

Code:
```
from sklearn.model_selection import train_test_split
X_train, X_test, y_train, y_test = train_test_split(X, y, test_size = 0.25)
```

Choosing a *K*-Value

Code:
```
errors = []
for i in range(1,100):
knn = KNeighborsClassifier(n_neighbors=i)
knn.fit(X_train,y_train)
pred_i = knn.predict(X_test)
```

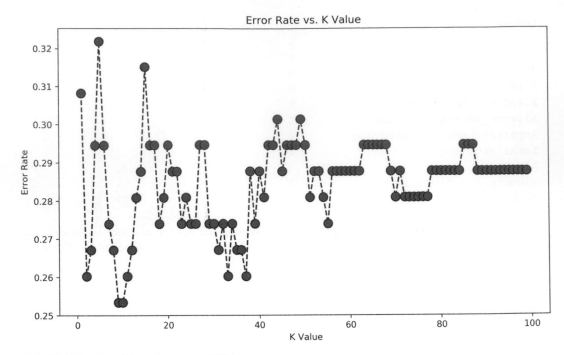

FIGURE 10.5 Plot of Error Rate versus *K*-Values.

```
errors.append(np.mean(pred_i!= y_test))
plt.figure(figsize = (10,6))
plt.plot(range(1,100),error_rate,color = 'blue', linestyle = 'dashed',
marker = 'o', markerfacecolor = 'red', markersize=10)
plt.title('Error Rate vs. K Value')
plt.xlabel('K Value')
plt.ylabel('Error Rate')
```

Output:
Based on Figure 10.5, it is deduced that "9" or "10" can be selected as a *k*-value, because they result in the least error. In this example, we will choose "9", because odd numbers have no chance of being in a tie while voting for a class.

Evaluation of the Model

Code:
```
from sklearn.neighbors import KNeighborsClassifier
classifier = KNeighborsClassifier(n_neighbors = 9)
classifier.fit(X_train, y_train)
y_pred = classifier.predict(X_test)
from sklearn.metrics import confusion_matrix, classification_report
print(confusion_matrix(y_test,y_pred))
print('\n')
print(classification_report(y_test,y_pred))
```

Output:
```
[[93 11]
 [26 16]]
```

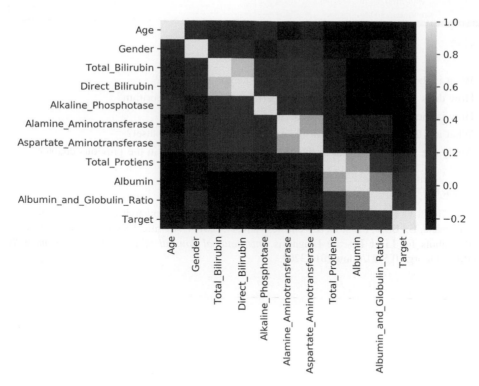

FIGURE 10.6 Heatmap of the Correlation Matrix for the Dataset.

	precision	recall	f1-score	support
1.0	0.78	0.89	0.83	104
2.0	0.59	0.38	0.46	42
accuracy			0.75	146
macro avg	0.69	0.64	0.65	146
weighted avg	0.73	0.75	0.73	146

The accuracy is 75%, and although this is higher than the base accuracy, the model did not perform significantly well. The factors resemble a high false negative. Low precision, recall, and the F1-score for class 2 suggested that the model is biased towards "1.0" or predicting as a liver patient. While class imbalance is the primary reason for this, there may be other reasons present, such as the low correlation of the features with the dependent variable. This can be verified by plotting the heatmap of the correlation matrix of this dataframe.

Code:

```
sns.heatmap(dataset.corr())
```

Output:

The intensities for the "Target" column show that it has very little correlation with all of the other variables, which may be between 0.2 and −0.2 (Figure 10.6). Therefore, the dataset is imbalanced and has a low correlation for the dependent variable - which are two of the major drawbacks for implementing the Machine Learning model on this dataset.

Having learned about the implementation of *K*-NN in Python, we shall cover two other supervised learning algorithms - decision trees and random forests - in the next chapter.

Exercise

1. What is the algorithm behind the *k*-nearest neighbor method?
2. How does the value of *"k"* affect the performance of this algorithm?
3. How is the value of *"k"* optimized?
4. What are the ways to deal with the missing attributes in a dataset?
5. Apply the *K*-NN algorithm to the breast cancer dataset used in the previous chapter.

REFERENCES

Dua, D., & Graff, C. (2017). *{UCI} Machine Learning Repository.* http://archive.ics.uci.edu/ml.

Venkata Ramana, B., Babu, M. S. P., & Venkateswarlu, N. (2011). A Critical Study of Selected Classification Algorithms for Liver Disease Diagnosis. *International Journal of Database Management Systems.* https://doi.org/10.5121/ijdms.2011.3207.

11

Decision Trees and Random Forests

Introduction

Decision trees and random forests are supervised learning algorithms that are used for problems of classification and regression. As the names suggest, random "forest" algorithms are essentially an assembly of decision "trees". A decision tree is based on the whole dataset, while a random forest incidentally selects observations and features to create several decision trees and then averages the results. In this chapter, we shall discuss both of these algorithms and their implementation in Python in detail.

A decision tree forecasts the value of a target variable by learning simple decision rules by using a tree-like hierarchical decision-making model. While this is a popular method that holds relevant applications in diverse real-life situations - such as civil planning, law, and business - it is also widely used in Machine Learning (Figure 11.1).

Decision trees work as a flow chart or a chain of decisions for achieving specific goals. illustrates the typical structure of a decision tree and its terminologies. Each node is a test or condition, and the branches represent their outcome.

- Root Node: This serves as the whole population or the sample that is split into two or more diverse sets of groups.
- Decision Node: The decision node is named when a sub-node divides into further sub-nodes.
- Leaf Node: Leaf nodes are the final outcomes of all of the respective chained decisions, and these are often called terminal nodes.

The process of dividing a node into further sub-nodes is based on a test that is called splitting.

How do decision trees train? Decision trees set test conditions based on information gain or the impurity in the dataset. In Figure 11.2, we have certain data points of two groups - blue and yellow - and carry two features X_1 and X_2 describing these data points. The first step is to find the condition of the root node. For the root node, we have to build a condition or a test through which we can separate the maximum data points of groups.

Accordingly, in step one, we selected the condition "$X_2 > 20$". As we can observe, the data points above and below the value "20" of the X_2 feature separates the entire population with minimum impurities, or specifically, the ratio of similar groups and the nonsimilar group is high (Figure 11.3). The root node (RN) has two branches - the left branch shows the "True" condition, and the right branch represents the "False" condition (i.e. "Yes" and "No" respectively, based on the root node's condition). In the second step, we repeat the same process for the subpopulations. We can notice that the data points can be subdivided for grouping similar data points based on feature X_1. For a data point, if the condition of the root node (RN) is false or X_2 feature <20, then we can find the X_1 feature. The X_1 feature can have a distinguished split for detaching the two groups at value "15". Therefore, in the right-hand side branch of the root node (RN), we can place a decision node (DN-1) bearing the condition $X_1 > 15$. Now, if a data point satisfies this condition, then it will belong to the yellow group, and if it fails, then the condition will belong to the blue group. Hence, these are the leaf nodes depicting outcomes.

Now, let us study the left branch of the root node (RN) where data points satisfy the condition of the root node. In this part, we can also subdivide the data points based on the X_1 feature following the rule of

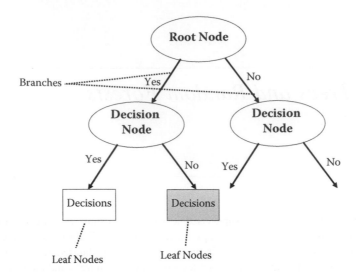

FIGURE 11.1   Decision Tree Structure.

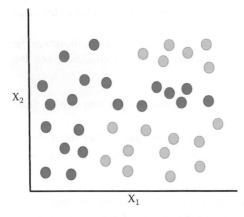

FIGURE 11.2 Example of Data Points for Decision Tree Classification.

having minimum impurity in the divided population. We can establish a decision boundary at $X_1 = 25$, where we can attain a pure class at $X_1 < 25$, hence, a leaf node. Therefore, we set the condition of this decision node (DN-2) as "$X_1 > 25$". If the condition fails, then we have a leaf node or else we have to execute another sub-division as illustrated by step three (Figure 11.4).

The left branch of DN-2 requires another decision boundary for the purification of data points. The decision will be based on the feature X_2, where we can acquire a decision boundary at $X_2 = 30$. Accordingly, after "DN-3", we have successfully built a decision tree to categorize the blue and yellow classes. Constructing too many decision nodes increases the depth of the decision tree, which leads to overfitting of the model. Therefore, with real-world data, we have to restrict the creation of decision nodes or boundaries while tolerating little impurities in leaf nodes. This process is called tree pruning.

Some of the main advantages of decision trees over other forms of supervised learning algorithms include, but are not limited to:

1. Decision trees are simple to understand and visually depictive.
2. Little to no data processing is required.
3. This can be applied for both numerical and categorical data.

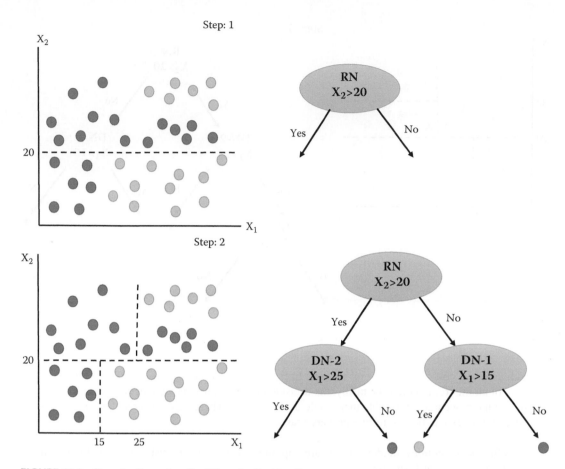

FIGURE 11.3 Steps for Generating Conditions for Decision Trees.

However, like any other Machine Learning algorithm, decision trees have their own set of disadvantages which include:

1. Decision trees are prone to overfitting. Methods such as pruning or specifying the minimum number of samples are used to overcome the problem of overfitting.
2. Since most of the decisions are made at individual nodes, the resultant tree may not be a globally optimal decision tree. Therefore, the dataset is balanced prior to fitting with the decision tree.
3. A single tree is sometimes insufficient to achieve successful results. The power of multiple decision trees is exploited to overcome this drawback of decision trees, whereby each node in the decision tree works on a random subset of features to calculate the output. This algorithm - also known as the random forest algorithm - then combines the output of individual decision trees to produce the end result.

Random Forests

As the name suggests, the random forest is a collection of decision trees. A random forest classifier trains several decision trees on small samples of the training dataset. These small samples are selected randomly with replacements, thereby constructing many decision trees. This technique of choosing small samples is called bootstrap sampling. The random forest model is also known as the ensemble

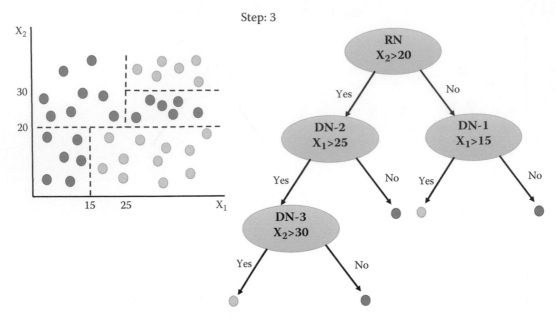

FIGURE 11.4   Step Three of Decision Tree Analogy.

method, as it is a collection of several decision trees. The prediction result is based on the cumulative outcome of all the trees, more specifically, the decision of the majority is the outcome. Usually, random forests are more accurate and stable than single decision trees. The greater accuracy of random forests over decision trees can be due to two primary reasons:

1. Unlike decision trees, random forests are unpruned, so the characteristic space is divided into a greater number of finer areas.
2. Each random forest tree learns from a random sample, and each node uses a random set of features to split-creating diversity among the trees.

Implementation of Decision Tree and Random Forest Using Sklearn

We will first try applying decision trees and the random forest on the heart disease dummy dataset. From previous chapters, we have learned that the base accuracy of this dataset is 54%. Let us see how well these two models perform.

Train Test Split

Code:

```
X = dataset.iloc[:,:-1]
y = dataset.iloc[:,-1]
from sklearn.model_selection import train_test_split
X_train, X_test, y_train, y_test = train_test_split(X, y, test_size=0.30)
```

Decision Trees

The class "DecisionTreeClassifier" is present in the "tree" model libraries of the sklearn package. We have already learned how to train a model using its "fit()" function.

Code:
```
from sklearn.tree import DecisionTreeClassifier
classifier = DecisionTreeClassifier()
classifier.fit(X_train,y_train)
```

Prediction and Evaluation

Code:
```
preds = classifier.predict(X_test)
from sklearn.metrics import classification_report,confusion_matrix
print(classification_report(y_test,preds))
```

Output:

	precision	recall	f1-score	support
0	0.82	0.82	0.82	44
1	0.83	0.83	0.83	47
accuracy			0.82	91
macro avg	0.82	0.82	0.82	91
weighted avg	0.82	0.82	0.82	91

The decision tree reached an accuracy rate of 82%, which is much higher than the baseline accuracy.

Predicting New Values

Code:
```
dataset.iloc[2,:]
```

Output:
```
age           41.0
sex            0.0
cp             1.0
trestbps     130.0
chol         204.0
fbs            0.0
restecg        0.0
thalach      172.0
exang          0.0
oldpeak        1.4
slope          2.0
ca             0.0
thal           2.0
target         1.0
Name: 2, dtype: float64
```

Code:
```
new_data=dataset.iloc[2,:-1].values
classifier.predict([new_data])
```

Output:
```
array([1], dtype=int64)
```

To conclude, the tree model has predicted the instance from the dataset correctly.

Random Forests

As we have discussed, the random forest classifier is an ensemble method; therefore, the "RandomForestClassifier" class is placed in the "ensemble" library of the sklearn package.

Code:
```
from sklearn.ensemble import RandomForestClassifier
classifier = RandomForestClassifier(n_estimators=200)
classifier.fit(X_train, y_train)
```

The "RandomForestClassifier" uses a parameter called the "n_estimators". During initiation, "n_estimators" is the number of trees to be generated for the training dataset. At this point, we input 200, meaning we it is our goal to create an ensemble of 200 decision trees.

Prediction and Evaluation of Random Forest Model

Code:
```
rfc_pred = classifier.predict(X_test)
print(confusion_matrix(y_test,rfc_pred))
print(classification_report(y_test,rfc_pred))
```
Output:
```
[[39 5]
 [ 8 39]]
              precision    recall  f1-score   support
           0       0.83      0.89      0.86        44
           1       0.89      0.83      0.86        47
    accuracy                           0.86        91
   macro avg       0.86      0.86      0.86        91
weighted avg       0.86      0.86      0.86        91
```
The random forest model achieved an accuracy rate of 86%, which is higher than the accuracy rate of 82% that was achieved with a single decision tree, thereby corroborating that the random forest is, in general, a better estimator than single decision trees.

Predicting Prognosis of Diabetes Using Random Forest

In this study, we will use an extended copy of the dataset of diabetes mellitus in a high-risk population of Pima Indian form (Smith et al., 1988). This data is available in the public domain and is also provided with the supplementary file. As always, the first step of a Machine Learning project is to load the dataset.

Loading Dataset

Code:
```
dataset = pd.read_csv('diabetes.csv')
dataset.head()
```

Output:
This dataset has nine features describing values for observation, and the last column labeled "Diabetic" represents the record stating if a person is diabetic, where "0" refers to a non-diabetic person and "1" refers to a diabetic person (Table 11.1).

Code:
```
dataset['Diabetic'].value_counts()
```

TABLE 11.1

Pima Indians Diabetes Dataset

	PatientID	Pregnancies	Plasma Glucose	Diastolic Blood Pressure	Triceps Thickness	Serum Insulin	BMI	Diabetes Pedigree	Age	Diabetic
0	1354778	0	171	80	34	23	43.509726	1.213191	21	0
1	1147438	8	92	93	47	36	21.240576	0.158365	23	0
2	1640031	7	115	47	52	35	41.511523	0.079019	23	0
3	1883350	9	103	78	25	304	29.582192	1.282870	43	1
4	1424119	1	85	59	27	35	42.604536	0.549542	22	0

Output:
```
0   10000
1    5000
Name: Diabetic, dtype: int64
```

The dataset has 10,000 observations for non-diabetics and 5,000 observations for diabetics. Therefore, the baseline accuracy is an estimate of 66.67%.

Code:
```
dataset.isnull().sum()
```

Output:
```
PatientID                0
Pregnancies              0
PlasmaGlucose            0
DiastolicBloodPressure   0
TricepsThickness         0
SerumInsulin             0
BMI                      0
DiabetesPedigree         0
Age                      0
Diabetic                 0
dtype: int64
```

To note, there are no missing values in this dataset.

Train-Test Split

Code:
```
X = dataset.iloc[:,1:-1]
y = dataset.iloc[:,-1]
from sklearn.model_selection import train_test_split
X_train, X_test, y_train, y_test = train_test_split(X, y, test_size = 0.25,
random_state = 0)
```

Training Classifier

Code:
```
from sklearn.ensemble import RandomForestClassifier
classifier = RandomForestClassifier(n_estimators=10)
classifier.fit(X_train, y_train)
```

In this part, we initiated this classifier with an ensemble of ten decision trees.

Code:
```
rfc_pred = classifier.predict(X_test)
print(classification_report(y_test,rfc_pred))
```

Output:

	precision	recall	f1-score	support
0	0.93	0.96	0.94	2500
1	0.91	0.85	0.88	1250
accuracy			0.92	3750
macro avg	0.92	0.90	0.91	3750
weighted avg	0.92	0.92	0.92	3750

The random forest classifier performed significantly well with this dataset, achieving an accuracy rate of 92%. Moreover, all of the values obtained for average precision, recall, and the F1 score were above 0.90.

Cross-Validation

While carrying out a Machine Learning project, we divide our dataset randomly into two sets - the training and the test set. However, there is a limitation to this technique. If we sacrifice around 20–25% of our data to evaluate our system, then our model may not be able to train properly for real-world challenges. Certain valuable information might get lost when we split the dataset. This will result in higher bias in our system. This problem gets aggravated when we work with a small dataset. To overcome this roadblock, we use a technique called cross-validation, as discussed in chapter 8. To do this, we split the dataset into various subsets and then train and evaluate our model with various combinations of these subsets. Sklearn's "model_selection" library has a "cross_validate" class to implement cross-validation.

Code:
```
from sklearn.model_selection import cross_validate
from sklearn.ensemble import RandomForestClassifier
classifier = RandomForestClassifier(n_estimators=10)
cross_validator = cross_validate(classifier, X, y, cv=5,
                                 scoring = 'accuracy',
                                 return_estimator = 'true')
print(cross_validator['test_score'])
```

Output:
```
[0.92633333   0.93533333   0.92733333   0.927   0.93633333]
```

The "cross_validate" class requires the "classifier", "features", "labels", and "cv", more precisely, the number of *k*-folds. We can establish a "return_estimator" parameter as true if it is our goal to retrieve the best estimator. We can observe in the output that five models have been trained with combinations of test and train data. The last model has the largest "test_score". We can selectta that model using the "best_model = cross_validator['estimator'][-1]" statement. The "best_model" is a random forest classifier object that can be used for future prediction.

Visualizing decision trees in Python requires additional installation of certain packages. An example of the corresponding implementation of visualizing decision trees has been shown in the Jupyter Notebook provided with this chapter in the supplementary files.

Exercise

1. How does the decision tree algorithm decide on the splitting condition for nodes?
2. Why is the random forest model a better estimator than simple decision trees?
3. Use data visualization techniques to understand the correlation between variables in the diabetes dataset.
4. Use a simple decision tree to classify the diabetes dataset and compare its performance with the random forest model.
5. What is cross-validation? How is it different from the standard train-test split method for training and evaluation of a model?

REFERENCE

Smith, J. W., Everhart, J. E., Dickson, W. C., Knowler, W. C., & Johannes, R. S. (1988). Using the ADAP learning algorithm to forecast the onset of diabetes mellitus. *Proceedings – Annual Symposium on Computer Applications in Medical Care*, 261–265.

12

Support Vector Machines

Introduction

Support vector machines, or SVMs, are a class of supervised Machine Learning algorithms and are one of the most favored for classification tasks, although these can also be used for regression (Wang & Lin, 2014). In this chapter, we will discuss the application of SVMs for classification tasks.

Support vector machines are linear classifiers, because they make linear decision boundaries. The support vector machine's goal is to find a hyperplane or decision boundary that is the best fit and separates n-dimensional space into separate classes or groups. When we place the new instance to a trained SVM, it classifies the new data point into one of the categories. The optimal decision boundary is known as a hyperplane. The hyperplane is a flat decision boundary with the dimensions $N - 1$ for an N-dimensional dataset. For 2D data, visually, it will be a line, and for 3D data, it will be represented with a plane separating two groups. For applying support vector machines, the data should be linearly classifiable, which means a line or plane should exist from where the groups or classes can be separated. The objective of SVM is to find that specific line of the hyperplane.

In Figure 12.1a we can observe that the dataset is linearly separable; however, there can be many lines that can separate the data into different classes. The support vector machine selects the optimal points that help in creating the hyperplane. The optimal points are the data points that are significantly similar but are categorized in different classes. SVM locates such points in the space and tries to accommodate a line or plane which has the maximum distance from those data points. These points are called support vectors, because they help in finding the decision boundary, and the algorithm is called a support vector machine.

SVM optimizes the margin of the classifier by using these support vectors, and any change in these support vectors would transform the hyperplane location. An example dataset is shown in Figure 12.1. In this figure, the margins are dividing the points neatly, and this margin is called the hard margin. With real-world data, we use a soft margin, which allows the SVM to misclassify a few points while maximizing the margin and minimizing the overall error. It is up to the user to decide the degree of tolerance or the "softness" of the margin. In sklearn's support vector machine class, this is controlled using a parameter is labeled "C". For the lower value of "C", the model will present less tolerance for misclassification, subsequently, for the higher value of "C" it will have a greater tolerance for misclassification

Kernel Trick

We have deliberated that SVMs can only classify linearly separable data, but datasets in the real-world may not always be linearly separable. Up to a certain extent, using a soft margin can help to classify data linearly, but, even then, the data must already possess a linear discriminatory boundary. For datasets that do not have any single linear boundary, the kernel trick is applied. This aims to map the low-dimensional data to a higher dimension where it is linearly separable. Let us understand this concept by using non-linearly separable one-dimensional data.

Figure 12.2a is a one-dimensional dataset that has two classes - yellow and blue. There is no way a point can differentiate these two classes. A one-dimensional dataset is a line only, and the hyperplane will be a point in this case, as a point has zero dimensions. Thus, the goal is to add an extra feature to the data so that we can differentiate the two classes using a line in two-dimensional space. In this way,

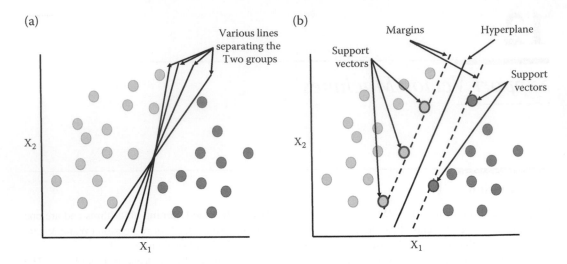

FIGURE 12.1   The Algorithm Behind Support Vector Machines.

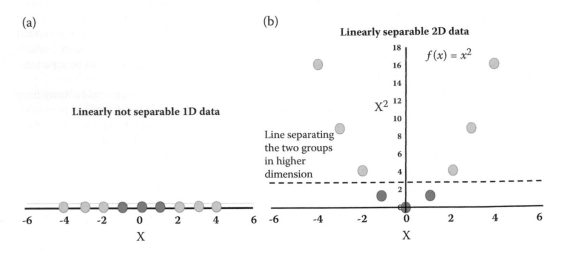

FIGURE 12.2   Mapping a Nonlinearly Separable One-Dimensional Data to High Dimension to Classify it Linearly.

the new feature is the square of the data points. Therefore, in Figure 12.2b, we can determine that the introduction of this new feature made the data points linearly separable into two different classes. Therefore, the assumption here is that, if N-dimensional data is not linearly separable, then there is a dimension greater than N where the data points are linearly separable. Kernal functions are the general functions that generate new features to map these data points to higher dimensions.

The radial basis function (RBF) is one of the common kernel functions used to perform the kernel trick. The "RBF" is used to create new higher dimensional features by calculating the distance among all of the other data points to a certain point. The mathematical representation of radial basis function is the following:

$$k(xi, xj) = exp(-\gamma \|x_i - x_j\|^2)$$

where x_i and x_j are two instances, and γ (gamma) controls the effect of new features on the decision boundary. Just as "C" is the regionalization parameter, gamma also needs to be tuned for the optimal performance of the SVMs.

Implementation of Support Vector Machines Using Sklearn

Let us try applying the support vector machine for the classification of the heart disease dataset.

Code:
```
dataset = pd.read_csv('heart.csv')
```

Train Test Split

Code:
```
X = dataset.iloc[:,:-1]
y = dataset.iloc[:,-1]
from sklearn.model_selection import train_test_split
X_train, X_test, y_train, y_test = train_test_split(X, y, test_size=0.30)
```

Train the Support Vector Classifier

The support vector classifier or the "SVC" class can be found in the "svm" library of the sklearn package. While the SVC class uses various parameters like "*c*" and "gamma" as described earlier in this chapter, we will first initialize the SVC class with default values and learn how it works. Next, we will proceed to parameter tuning.

Code:
```
from sklearn.svm import SVC
model = SVC()
model.fit(X_train,y_train)
```

Predictions and Evaluations

Now, let us forecast the test set using the trained model.

Code:
```
preds = model.predict(X_test)
from sklearn.metrics import classification_report,confusion_matrix
print(confusion_matrix(y_test,preds))
print('\n')
print(classification_report(y_test,preds))
```

Output:
```
[[20 24]
 [ 6 41]]
```

	precision	recall	f1-score	support
0	0.77	0.45	0.57	44
1	0.63	0.87	0.73	47
accuracy			0.67	91
macro avg	0.70	0.66	0.65	91
weighted avg	0.70	0.67	0.65	91

With default settings, the support vector classifier achieved an accuracy rate of 67%, which is better than the baseline accuracy of our dummy dataset, specifically, the heart disease dataset. However, this can be improved by tuning the values of "*c*" and "gamma".

Grid Search

We have discussed that the "*c*" value provides flexibility to the margins of an SVM and provides it with tolerance towards some error. However, the question of the amount of tolerance we can afford for our support vector class so that it can provide satisfactory accuracy has no straightforward answer. In the same way, choosing the value "gamma", which defines the effect of new features while increasing the dimensions using a radial basis function, depends on a case-to-case basis. While the radial basis function is the most commonly used kernel function, there are also other functions - such as polynomial and sigmoidal. Therefore, selecting the correct kernel function is also a challenging task. The solution to all of these obstacles is known as hyperparameter tuning, as discussed in chapter 8. In this process, "*c*", "gamma", and kernel functions are the parameters. We are required to determine the optimal combination of all of these parameters. The process of selecting the optimal pairs of parameters from a combination of various pairs is called a grid search. In the grid search, we check all of the combinations of parameters and choose the best performing one. A grid search is a very time-consuming process if we have a lot of parameters along with their values. However, it is a popular way to find ideal parameters. It is significantly well known among data scientists that sklearn has an estimator class called "GridSearchCV" in its model selection library. "GridSearchCV" acts like any Machine Learning parameter, except its usage on a dictionary of parameter space in which we prefer to run the grid search for the classifier class.

Code:
```
from sklearn.model_selection import GridSearchCV
param_grid = {'C': [0.1,1, 10, 100, 1000], 'gamma': [1,0.1,0.01,0.001,0.0001],
'kernel': ['rbf']}
grid = GridSearchCV(SVC(),param_grid,verbose=3)
grid.fit(X_train,y_train)
```

In the block of code above, we created a dictionary of parameters called the "param_grid" where we provided various values of "*c*" and "gamma". A grid search takes a significant amount of time as it trains many models with all the combinations of parameters. At this point, we are selecting "RBF" as the only kernel function to save time. Next, we fit our training set to the "grid" object, as it behaves like a Machine Learning estimator.

Code:
```
grid.best_params_
```

Output:
```
{'C': 100, 'gamma': 0.0001, 'kernel': 'rbf'}
```

After the grid search is completed, we can retrieve the best parameters using the "best_parms_" attribute of a GridSearchCV object. Here we attained "*c*" = 100 and "gamma" = 0.0001 as an optimal combination with the "RBF" kernel. The GridSearchCV object can be used to predict, as it will have all the properties of the classifier with the best parameters. Now, let us evaluate this hyperparameter tuned classifier using the test set.

Code:
```
grid_predictions = grid.predict(X_test)
print(confusion_matrix(y_test,grid_predictions))
print('\n')
print(classification_report(y_test,grid_predictions))
```

Output:
```
[[32 12]
 [ 4 43]]
              precision   recall  f1-score  support
         0         0.89     0.73      0.80       44
         1         0.78     0.91      0.84       47
  accuracy                           0.82       91
 macro avg         0.84     0.82      0.82       91
weighted avg       0.83     0.82      0.82       91
```

As expected, the tuned model has compellingly increased accuracy. The support vector machines are rather sensitive to their parameters, so it is extremely important to find optimal parameters for every SVM project.

Prediction of Wheat Species Based on Wheat Seed Data

In this project, we will use the attributes of wheat seed for the prediction of their species. This data was first used in a study by (Charytanowicz et al., 2010), and the data is publicly available at the UCI Machine Learning Repository (Dua & Graff, 2017). The dataset is comprised of seven independent features and 210 observations.

Code:
```
dataset = pd.read_csv('seeds_dataset.csv')
dataset.head()
```

Output:
In this example, the last column contains the seed's species labels. We will omit the first column or the "ID" while creating independent features (Table 12.1).

Code:
```
dataset['seedType'].value_counts()
```

Output:
```
3    70
2    70
1    70
Name: seedType, dtype: int64
```
The output shows that we have three classes, and all have an equal distribution of instances, specifically 70 for each group. Here, the baseline accuracy is 33%.

TABLE 12.1

Attributes of the Wheat Seed Dataset

	ID	Area	Perimeter	Compactness	Length of Kernel	Width of Kernel	Asymmetry Coefficient	Length of Kernel Groove	Seed Type
0	1	15.26	14.84	0.8710	5.763	3.312	2.221	5.220	1
1	2	14.88	14.57	0.8811	5.554	3.333	1.018	4.956	1
2	3	14.29	14.09	0.9050	5.291	3.337	2.699	4.825	1
3	4	13.84	13.94	0.8955	5.324	3.379	2.259	4.805	1
4	5	16.14	14.99	0.9034	5.658	3.562	1.355	5.175	1

Code:
```
dataset.isnull().sum()
```

Output:
```
ID                          0
area                        0
perimeter                   0
compactness                 0
lengthOfKernel              0
widthOfKernel               0
asymmetryCoefficient        0
lengthOfKernelGroove        0
seedType                    0
dtype: int64
```

To note, the dataset contains no null values.

Train Test Split

Code:
```
X = dataset.iloc[:,1:-1]
y = dataset.iloc[:,-1]
from sklearn.model_selection import train_test_split
X_train, X_test, y_train, y_test = train_test_split(X, y, test_size = 0.25,
random_state = 0)
```

Traning Support Vector Classifier and Tuning Its Parameters Using a Grid Search

In this part, the search parameters are defined in the "param_grid" dictionary, and the "GridSearchCV" object is fitted using a training set.

Code:
```
from sklearn.model_selection import GridSearchCV
from sklearn.svm import SVC
param_grid = {'C': [0.1,1, 10, 100, 1000], 'gamma':
[1,0.1,0.01,0.001,0.0001], 'kernel': ['rbf']}
grid = GridSearchCV(SVC(),param_grid,refit=True,verbose=3)
grid.fit(X_train,y_train)
```

Code:
```
grid.best_params_
```

Output:
```
{'C': 1000, 'gamma': 0.1, 'kernel': 'rbf'}
```

The output presents the best parameters for the support vector classifier for this dataset.

Code:
```
grid_predictions = grid.predict(X_test)
print(classification_report(y_test,grid_predictions))
```

Output:

	precision	recall	f1-score	support
1	0.89	1.00	0.94	17
2	1.00	0.90	0.95	21
3	1.00	1.00	1.00	15
accuracy			0.96	53
macro avg	0.96	0.97	0.96	53
weighted avg	0.97	0.96	0.96	53

We attained an accuracy rate of 96%, thereby demonstrating that support vector machines are excellent classifiers, because they find an optimal hyperplane between different classes where data points are significantly similar to one another (i.e. the support vectors). In contrast, most other classification algorithms operate on the principle of finding similarity within the classes. This property of SVM makes them less sensitive to outliers, and they only consider the support vectors. The limitation of SVMs is that they are linear model and require high computation power for operating on non-linearly separable datasets, since they have to be mapped to higher dimensions. In the next chapter, we will discuss artificial neural networks.

Exercise

1. How do support vector machines classify linearly separable data?
2. What is the role of the parameter "gamma" in radial basis function?
3. How does the grid search optimize the parameters of a Machine Learning model?
4. Apply SVM on the breast cancer and diabetes datasets and compare the performance with other classification algorithms.

REFERENCES

Charytanowicz, M., Niewczas, J., Kulczycki, P., Kowalski, P. A., Łukasik, S., & Zak, S. (2010). Complete gradient clustering algorithm for features analysis of X-ray images. *Advances in Intelligent and Soft Computing*. https://doi.org/10.1007/978-3-642-13105-9_2.

Dua, D., & Graff, C. (2017). *{UCI} Machine Learning Repository*. http://archive.ics.uci.edu/ml.

Wang, P. W., & Lin, C. J. (2014). Support vector machines. In *Data Classification: Algorithms and Applications*. https://doi.org/10.1201/b17320.

13

Neural Nets and Deep Learning

Introduction

After studying the different types of Machine Learning algorithms and their applications, we shall study another widely used supervised learning algorithm, artificial neural networks, ANNs, or neural nets in this chapter. As the name suggests, ANNs were developed to simulate the neural networks or neurons that constitute a human brain. They are mostly used for statistical investigation and modeling of gathered data. Their role is viewed as a substitute to standard nonlinear regression analysis models. They are characteristically employed in unraveling problems that may be expressed in terms of prediction or classification. With around six decades of research, neural networks find their applications in diverse fields - ranging from speech and image recognition and classification, text recognition, medical diagnosis, and fraud detection, among others.

Neural Networks Architecture

Figure 13.1 illustrates the general architecture of neural networks. It consists of three distinct layers: the input layer, the hidden layer, and the output layer. The foremost layer is the input layer that houses the input features or input neurons. The middle layer is the hidden layer; the term "hidden" implies the processes of mathematical computation that are not easily visible and are sometimes also termed as the black box. Diverse networks are characterized by numbers of hidden layers based on application. An

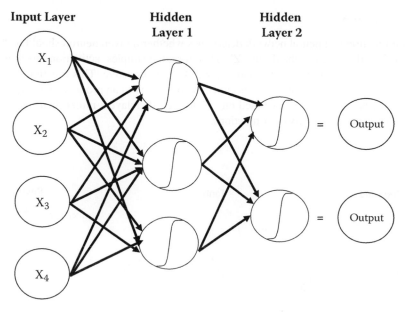

FIGURE 13.1   Simple Architecture of Artificial Neural Networks.

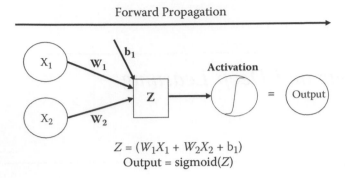

$$Z = (W_1X_1 + W_2X_2 + b_1)$$
$$Output = sigmoid(Z)$$

FIGURE 13.2 Forward Propagation, Calculating the Output.

ANN consisting of more than one hidden layer is called a deep neural network. The last layer is the output layer, which contains the output of the network.

The Working Principle of Neural Networks

We shall first use a simple neuron as an example and study how it calculates or predicts output.

Neuronsprimarily consist of inputs, features, weights, bias, and activation function. Weights define the importance of features, just as we have observed in the linear regression model. The sum of weighted features are passed to an activation function and are calculated using the following formula:

$$Z = W_1X_1 + W_2X_2 + b$$

Where the "Ws" are the weights for the feature, "Xs" are the input features, and "b" is the bias. Bias is a small random number and is used so that "Z" does not become equal to zero for any values of Ws and Xs (Figure 13.2).

Activation Functions

The activation function in a neural network determines whether a given neuron should be "activated" or "not activated" based on the weighted sum "Z". Therefore, the simplest activation function will be $f(z) = 1$, if "z" is greater than zero or positive, and $f(z) = 0$ if "z" is negative. This function is called the step function because it provides an output that is either "0" or "1". We can produce other types of activation functions that can pass certain values based on "Z" to the other layers rather than only "1" or "0".

Figure 13.3 presents three activation functions:

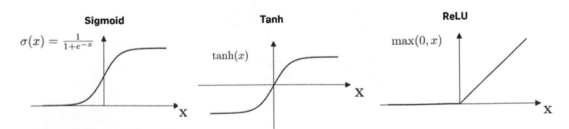

FIGURE 13.3 Activation Functions.

1. The "sigmoid" function returns a value between "0" and "1" for any value of "z". The sigmoid function is generally used in the output layer of a deep neural network.
2. The "Tanh" function will produce output ranging from "-1" to "1", when "z" is passed through it.
3. The "ReLU" function, also known as the rectified linear unit, is the most commonly used activation function. It delivers "0" for negative values of "z" and will return the "z" if it is a positive value.

Steps of Forward Propagation

Let us revisit Figure 13.2 and note the steps for the calculation of output based on input features:

1. Input features are assigned weights.
2. The weighted sum of all of the input features with the addition of bias is calculated (i.e. "Z").
3. The value of Z is passed through the activation function and returns the output.

The main objective of neural networks is to calculate appropriate weights for the input features to obtain an output. The output is compared with the original values, and an error or loss is calculated. Next, this loss is driven backwards to optimize the weights. This process of calculation of output, comparing loss, and updating the weights runs iteratively until we attain a minimum loss. This process of minimizing the error is called gradient descent.

Gradient Descent

The gradient descent algorithm determines how the weight is updated. Supposing we have an output calculated through the forward propagation method "$\hat{y}$", and the actual value is "y". At this point, we will calculate the error using a function, and this function is called the loss function or cost function. Means square error, which is the sum of all of the squared errors as we have discussed in chapter 8, is also a type of loss or cost function. A single output value will be computed in this manner:

$$Loss = 1/2 * (y - \hat{y})^2$$

If the output is calculated using forward propagation steps for "n" number of times with different random weights while minimizing the error, then we will obtain a graph as shown in Figure 14.4 when plotting loss versus weight. Next, we can select the weight in which the error is at the minimum. These steps are feasible with one or two features, two weights, and one neuron. However, with many features, weights, and neurons, finding the best combination of weights will take an indefinite amount of time when following the above-mentioned process. Therefore, the gradient descent algorithm has evolved as an alternative method - wherein the weights are updated in such a way that, every time, the loss is minimized, and the model thus obtained will procure minimum error (Figure 13.4).

The basic idea behind the gradient descent algorithm is that differentiating a function provides its derivative - which is the slope for the function, more specifically called the gradient. Once we have

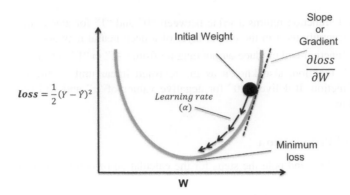

FIGURE 13.4   Gradient Descent.

$$W_{new} = W - \alpha * \frac{\partial loss}{\partial W}$$ **FIGURE 13.5**   Formula of Updating Weight.

acquired the gradient, we can proceed with the minimum value of the function. The famous hilltop example can further help in understanding this. Supposing we are somewhere at the top of a hill, and we want to go downside of the hill while blindfolded. Although we cannot see the bottom of the hill, we can feel the slope with our feet, and we are able to take steps towards the bottom of the hill. Therefore, we can locate the direction of minimum error by differentiating the loss function in terms of the weights and backpropagate this information for updating the weight (Figure 13.5). The amount of weight that should be updated depends on the derivative of the loss function or the gradient and learning rate (α).

The user defines the learning rate, and it is the length of the step the algorithm takes to reach the minimum loss value. If the step is too long then it might miss the minimum value. Subsequently, if the step is too short, then it will take more time to reach the bottom. There are many variants of gradient descent algorithm - such as RMSprop, adam, etc. - and these are called optimizers because they enhance the weights and, hence, the predictions.

Backpropagation

Through the process of forward propagation, we understand that the output of the activation "A" is the function of "Z"; "Z" is the function of weight "W"; the bias "b". To calculate the derivative of loss in terms of weight, we use partial derivatives of each function. First, we compute the derivative of the loss function in terms of the activation function. Second, we use this value to find the derivative of the activation function in terms of "Z". Lastly, we determine the derivative of "Z" in

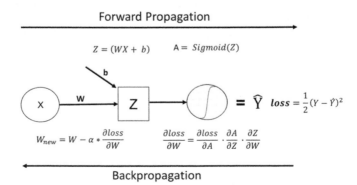

FIGURE 13.6   Forward and Backward Propagation for Neural Networks.

terms of "*W*". This process of propagating the loss backwards for updating the weights is called backpropagation.

In a deep neural network, this becomes complex while we follow these forward propagation and backpropagation processes iteratively for updating each weight. Moreover, when the number of layers is more than one, the derivative chain becomes longer (Figure 13.6), but the idea remains the same.

Having a number of weights and activation functions provides neural networks with the ability to handle linear as well as nonlinear datasets. These features of neural networks also make them susceptible to overfitting the training data. Therefore, they require a significant amount of parameter tuning to obtain optimal results.

Implementing Neural Networks Using TensorFlow

As we have studied so far, training of neural networks contains many functions and mathematics; however, Python packages deliver their implementation very straight forward. There are various packages available in Python for carrying out neural networks. These include TensorFlow, Pytorch, and Theano, among others. We will use TensorFlow, which is an open-source deep learning library developed by Google. TensorFlow has two types of packages. The first type is the normal TensorFlow which uses a CPU for calculations, and the other is the GPU version. As neural networks involve a significant amount of computations to be carried out simultaneously, the GPU version shifts calculations to the graphics card for faster processing. For the project in this book, any version can be used as we will be handling small datasets, but the GPU version is highly recommended when handling large datasets. First, we will use the heart disease dataset for classification (Table 13.1).

Code:

```
dataset = pd.read_csv('heart.csv')
dataset.head()
```

Output:
We then split the dependent and independent variables.

Code:

```
X = dataset.iloc[:,:-1]
y = dataset.iloc[:,-1]
```

TABLE 13.1

The Heart Disease Dataset

	Age	Sex	cp	trestbps	chol	fbs	restecg	thalach	exang	Oldpeak	Slope	ca	thal	Target
0	63	1	3	145	233	1	0	150	0	2.3	0	0	1	1
1	37	1	2	130	250	0	1	187	0	3.5	0	0	2	1
2	41	0	1	130	204	0	0	172	0	1.4	2	0	2	1
3	56	1	1	120	236	0	1	178	0	0.8	2	0	2	1
4	57	0	0	120	354	0	1	163	1	0.6	2	0	2	1

Data Scaling

Data scaling is an essential preprocessing stage set before training a neural network, as large variations in parameter distributions can affect the activation functions.

Code:

```
from sklearn.preprocessing import StandardScaler
sc = StandardScaler()
X_norm = sc.fit_transform(X)
from sklearn.model_selection import train_test_split
X_train, X_test, y_train, y_test = train_test_split(X_norm, y, test_size =
0.25, random_state = 0)
```

TensorFlow 2.0

In this example, TensorFlow Version 2.0 is used - which is the current stable version. This version of TensorFlow can be retrieved using the following code:

Code:

```
import tensorflow as tf
tf.__version__
```

Output

```
'2.0.0'
```

TensorFlow has a library called Keras, which provides simple initiative APIs for building neural networks. It has a class called "Sequential" which can be used to create layers of neurons. We can use "help(<Class_Name>)" to locate details about any class in the Jupyter Notebook. Now, let us import this class and built certain models.

Code:

```
import tensorflow as tf
from tensorflow.keras.models import Sequential
from tensorflow.keras.layers import Dense
```

Dense class is used to build densely connected neural networks, where every neuron is connected to every other neuron.

Creating a Model

There are two methods for the creation of multiple layer neural networks using a sequential class. First is passing a list of layers all at once, and the second is adding layers one by one.

Model – As a List of Layers

Code:

```
model = Sequential([
      Dense(units=2),
      Dense(units=2),
      Dense(units=1)
])
```

In this part, we initialize the class "Sequential" and pass a list of "Dense" objects. "Dense" uses various parameters, such as "units", which indicate the number of neurons required in the respective layer. We will discuss other parameters as we come across them in this project. Now, we created three layers of the neural network and saved it as an object in the "model" variable.

Model – Adding in Layers One by One

Furthermore, we can also add dense layers one by one to the model object by using the ".add()" method. The model is defined in the block of code above, and the block of code below is similar.

Code:
```
model = Sequential()

model.add(Dense(2))
model.add(Dense(2))
model.add(Dense(1))
```

Building Model

In this part, we will build a model to classify the heart disease dataset.

Code:
```
import tensorflow as tf
from tensorflow.keras.models import Sequential
from tensorflow.keras.layers import Dense
model = Sequential()

#
model.add(Dense(units=13,activation='relu',input_shape=(13,)))
model.add(Dense(units=7,activation='relu'))

model.add(Dense(units=1,activation='sigmoid'))

# For a binary classification problem
model.compile(loss='binary_crossentropy',optimizer='adam',
                              metrics=["accuracy"])
model.summary()
```

Output:
```
Model: "sequential"
```

Layer (type)	Output Shape	Param #
dense (Dense)	(None, 13)	182
dense_1 (Dense)	(None, 7)	98
dense_2 (Dense)	(None, 1)	8

```
Total params: 288
Trainable params: 288
Non-trainable params: 0
```

The dense object uses the number of units and the activation function for the respective layer. In this case, we selected "relu" or the rectified linear unit activation function. It is important to note that, for the first layer, there is an optional parameter named "input_shape"; it adopts the dimension of input features as an input, and we have "13" features here. For the last layer, we have selected the "sigmoid" activation function, since the problem under consideration is a binary classification problem where we have only two classes. There is no definite rule of thumb to follow when determining the number of layers and the number of neurons in each layer. However, the best practice to calculate the number of neurons is rounding of half of the number of features plus the number of output neurons, because the number of neurons cannot be a fraction. In this example, the features are "13", and the output has "1" neuron. Therefore, we selected "7" neurons for the hidden layer. After adding the layers, we compiled the model. While compiling the model, we declare the parameters to be depicted while training the model - such as the loss function, optimizer, and the evaluation matrix. Different loss functions are used for a different task, and this also depends on the number of neurons in the last layer. Table 13.2 shows some satisfactory combinations of these parameters in different Machine Learning tasks.

Since this task is a binary classification problem, we will use "binary_crossentropy" as the loss function; the role of the optimizer is to minimize this loss. "model.summary()" illustrates the architecture of the model in the output.

Training Model

Code:

```
model.fit(x=X_train,
          y=y_train,
          epochs=400,
          validation_data=(X_test, y_test), verbose=1
          )
```

Output:

```
Train on 227 samples, validate on 76 samples
Epoch 1/400
227/227 [==============================] - 4s 17ms/sample - loss:
0.8407 - accuracy: 0.4405 - val_loss: 0.9136 - val_accuracy: 0.3289
 .
 .
 .

Epoch 400/400
227/227 [==============================] - 0s 547us/sample - loss:
0.0994 - accuracy: 0.9692 - val_loss: 0.4028 - val_accuracy: 0.8816
```

TABLE 13.2

Loss Functions for Various Machine Learning Tasks

Machine Learning Task	Loss Functions	Optimizer
Multi-class classification	'categorical_crossentropy'	'rmsprop', 'adam'
Binary classification	'binary_crossentropy'	'rmsprop', 'adam'
Regression problem	'mse'	'rmsprop', 'adam'

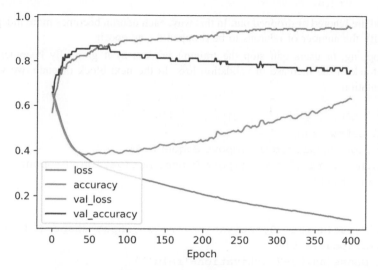

FIGURE 13.7   Training versus Validation Performance of the Model.

As apparent in sklearn objects, the Keras sequential object also has the "fit" method for training the model. The "fit" method takes the training variables, the test variables, and the epochs. Epochs are the number of times we want to iterate forward and backward propagation on whole training data. We have learned that with each iteration, the loss will be minimized. However, too many iterations will overfit the training data. To overcome this problem, we will first do a trial with the method to check if the model is overfitted. Next, we will learn how to prevent overfitting. The "fit" method also uses "verbose" to get the visually attractive output after the completion of each epoch while training the model.

Overfitting

The performance parameter of the trained model can be found in the "history.histroy" attribute of the sequential object. It contains losses and accuracy, as we passed "accuracy" as the matrix.

Code:
```
model_performance = pd.DataFrame(model.history.history)
ax = model_performance.plot()
ax.set_xlabel('Epoch')
```

Output:
Figure 13.7 shows that for initial epochs, the training and validation accuracy is almost equal. With a few more epochs, the "val_accuracy" becomes less than the training accuracy. This is the sign of overfitting in training data. The same trend can be observed for the loss, because the val_loss starts to increase after certain epochs. After 400 epochs, our model is overfitted.

Dropout and Early Stopping

The two common methods used to reduce overfitting are:

1. **Adding dropout layers to the model, or when the** completely connected layer optimizes all of the weights of every neuron and, thus, neurons establish codependence among them. This reduces the actual power of each neuron, thereby resulting in overfitting the training data. This can be addressed by silencing some neurons during training. Neurons are selected randomly at each

epoch and are ignored or dropped out. In this way, each neuron becomes more independent. Users can specify the number of neurons to be dropped for an epoch.

2.. Early stopping, or users can stop the training process as it gradually becomes overfitted by keeping track of the increase in validation loss. In the next block of code, we will study their implementation.

Code:

```
import tensorflow as tf
from tensorflow.keras.models import Sequential
from tensorflow.keras.layers import Dense, Activation, Dropout
model = Sequential()

#
model.add(Dense(units=13,activation='relu',input_shape=(13,)))
model.add(Dropout(0.5))
model.add(Dense(units=7,activation='relu'))

model.add(Dropout(0.5))
model.add(Dense(units1,activation='sigmoid'))

# For a binary classification problem
model.compile(loss='binary_crossentropy',
optimizer='adam',metrics=["accuracy"])
model.summary()
```

Output:

```
Model: "sequential_1"
```

Layer (type)	Output Shape	Param #
dense_3 (Dense)	(None, 13)	182
dropout (Dropout)	(None, 13)	0
dense_4 (Dense)	(None, 7)	98
dropout_1 (Dropout)	(None, 7)	0
dense_5 (Dense)	(None, 1)	8

```
Total params: 288
Trainable params: 288
Non-trainable params: 0
```

The dropout layer is introduced between the layers of densely connected neurons. The dropout layers takes a value between "0" and "1", denoting the fraction of the neurons that users have opted to drop out at each epoch.

Code:
```
from tensorflow.keras.callbacks import EarlyStopping
early_stop = EarlyStopping(monitor='val_loss', mode='min', verbose=1,
patience=5)
model.fit(x=X_train,
          y=y_train,
          epochs=400,
          validation_data=(X_test, y_test), verbose=1,
          callbacks=[early_stop]
          )
```

Output:
```
Train on 227 samples, validate on 76 samples
Epoch 1/400
227/227 [==============================] - 2s 9ms/sample - loss: 0.6937 -
accuracy: 0.5991 - val_loss: 0.7043 - val_accuracy: 0.5263
.
.
.
.
Epoch 152/400
227/227 [==============================] - 0s 655us/sample - loss: 0.4853 -
accuracy: 0.7665 - val_loss: 0.4106 - val_accuracy: 0.8026
Epoch 00152: early stopping
```

An early stopping object is passed to the fit method as a callback object. Callback objects can perform tasks during the training of the model, such as blocking the model while keeping track of the validation in the case of the "EarlyStopping" callback. The "EarlyStopping" class uses an argument to monitor; the "mode" parameter indicates the type of tracking, such as if we want to track validation loss, then it is our goal for it to be minimum. In case we want to track validation accuracy, we will retain as "max". The "patience" parameter is for standing by for some epochs after the condition is satisfied, as the curve of loss or accuracy possess spikes. Therefore, there could be certain gains or losses, and we prefer our early stopping object to ignore such intermittent spikes. In the output, we can observe that the model stopped after 152 epochs. With early stopping, we can put a large number of epochs, as the model will stop training automatically the moment it starts overfitting.

Code:
```
model_performance = pd.DataFrame(model.history.history)
ax = model_performance.plot()
ax.set_xlabel('Epoch')
```

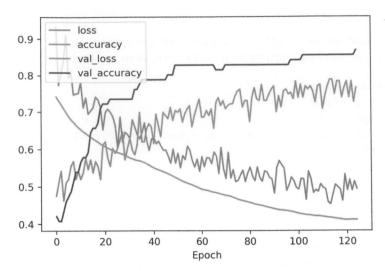

FIGURE 13.8   Plot of Models Training versus Validation Performance after Using the Dropout Layer and Early Stopping Object.

Output:

In Figure 13.8, we can observe that the model training is terminated before overfitting - more specifically, validation loss is below the training loss - and validation accuracy is above the training accuracy.

Model Evaluation

The model object has a method called "predict_classes", where we can pass the test set for prediction. We will use sklearn's classification_report and confusion_matrix for the evaluation of the model.

Code:

```
predictions = model.predict_classes(X_test)

from sklearn.metrics import classification_report,confusion_matrix

print(confusion_matrix(y_test,predictions))
print(classification_report(y_test,predictions))
```

Output:

```
[[20 13]
 [ 2 41]]
```

	precision	recall	f1-score	support
0	0.91	0.61	0.73	33
1	0.76	0.95	0.85	43
accuracy			0.80	76
macro avg	0.83	0.78	0.79	76
weighted avg	0.82	0.80	0.79	76

The model has an 80% accuracy rate which is much higher than the base accuracy rate of this model - which was 56%.

Predicting New Instance

The "predict" method of this model produces the probability of classification. Before the forecast, the new data has to be normalized through the "sc" object, which is used to transform the independent features.

Code:
```
new_data=dataset.iloc[2,:-1].values
model.predict(sc.transform([new_data]))
```

Output:
```
array([[0.9287414]], dtype=float32)
```

In this case, we attained the probability as 0.93 of belonging to class "1". We can produce the class using the "predict_classes" method.

Predicting Breast Cancer Using Neural Networks

In this project, we will use the Breast Cancer Wisconsin (Diagnostic) Dataset, which we used in chapter 9 on Logistic Regression. This dataset contains 30 features that are derived from cell images of benign and malignant breast tumors. There are 569 observations - where 357 instances are benign, and 212 instances are malignant tumors. Therefore, the baseline accuracy of the model is 62.7%. There were no "null" values, but the dataset contained two undesired columns, (i.e. "id" and "Unnamed:32") which will be removed. There are also classes are in string datatype, so we will convert these into numeric type by labeling "B" as "0", and "M" as "1".

Code:
```
dataset = pd.read_csv('Breast_Cancer.csv')

dataset.drop(['id','Unnamed: 32'],axis=1,inplace=True)
dataset['diagnosis'] = dataset['diagnosis'].map({'B':0,'M':1})
```

Separating the Dependent and Independent Dataset

Code:
```
X = dataset.iloc[:,1:].values
y =dataset.iloc[:,0].values
```

Data Scaling

Code:
```
from sklearn.preprocessing import StandardScaler
sc = StandardScaler()
X_norm = sc.fit_transform(X)
```

Splitting the Dataset into the Training Set and Test Set

Code:
```
from sklearn.model_selection import train_test_split
X_train, X_test, y_train, y_test = train_test_split(X, y, test_size = 0.20,
random_state = 101)
```

Creating the Model

At this point, we will build three layers. The first layer has "30" neurons; the second layer has "15" neurons; the last layer has a single neuron with a sigmoid activation function, since it is a binary classification problem. In between, we have added dropout layers with the probability of dropping out neurons indicated as "0.5". While compiling the model, we used "binary_crossentropy" as a loss function and "rmsprop" as the optimizer. Lastly, we fitted the model with the early stopping callback object, which will prevent the model from overfitting.

Code:

```
import tensorflow as tf
from tensorflow.keras.models import Sequential
from tensorflow.keras.layers import Dense, Activation, Dropout
model = Sequential()

#
model = Sequential()
model.add(Dense(units=30,activation='relu'))
model.add(Dropout(0.5))

model.add(Dense(units=15,activation='relu'))
model.add(Dropout(0.5))

model.add(Dense(units=1,activation='sigmoid'))
model.compile(loss='binary_crossentropy',optimizer='rmsprop',
                      metrics=['accuracy'])

from tensorflow.keras.callbacks import EarlyStopping
early_stop = EarlyStopping(monitor='val_loss', mode='min', verbose=1,
patience=25)
model.fit(x=X_train,
          y=y_train,
          epochs=400,
          batch_size= 64,
          validation_data=(X_test, y_test), verbose=1,
          callbacks=[early_stop]
)
```

Next, we will plot the validation loss and training loss data to confirm if the model is overfitted or not.

Code:

```
model_loss = pd.DataFrame(model.history.history)
ax = model_loss[['loss','val_loss']].plot()
ax.set_xlabel('Epoch')
```
Output:

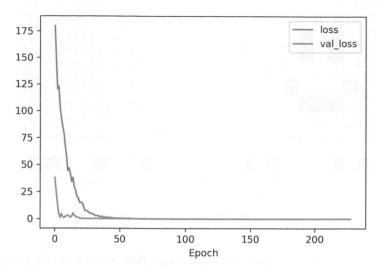

FIGURE 13.9 Training Loss versus Validation Loss Plot.

The plot in Figure 13.9 illustrates that the training and validation loss are almost similar, implying that the model is not overfitted.

Model Evaluation

Code:

```
predictions = model.predict_classes(X_test)
from sklearn.metrics import classification_report,confusion_matrix

print(confusion_matrix(y_test,predictions))
print(classification_report(y_test,predictions))
```

Output:

```
[[72  0]
 [ 9 33]]
              precision    recall  f1-score   support
           0       0.89      1.00      0.94        72
           1       1.00      0.79      0.88        42
    accuracy                           0.92       114
   macro avg       0.94      0.89      0.91       114
weighted avg       0.93      0.92      0.92       114
```

The model achieved a 92% accuracy rate with a recall of "1" for the benign class, or more precisely, no wrong prediction for the benign class.

In the next section, we will study a special type of neural network which is used primarily with image data.

Convolutional Neural Network

For computers, images are a collection of pixel intensities stored in a specific order. For us, the image is a collection of colors. We have seen image puzzles, where images are chopped into small subparts, and then the succeeding task is to rejoin those small parts to form a full image. The subparts alone do not have any meaning. This is the case with simple neural networks as well.

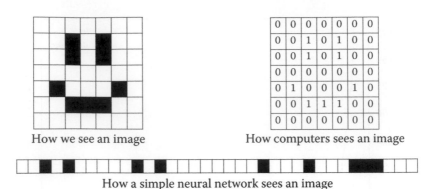

How we see an image   How computers sees an image

How a simple neural network sees an image

FIGURE 13.10 Image Data with Various Prospectives.

If we want to train an artificial neural network in image data, we have to plug out all of the pixels from the image and feed each pixel into the neural network as a feature. Although individual features have information about the image, images are more understandable when local pixels are assembled together. In Figure 13.10, we can see how an image becomes distorted when it is fed into neural networks.

Convolution neural networks (CNNs) are well-known deep learning models for image data or spatial data. CNNs can learn local spatial features, while ANNs can learn features individually. A CNN uses a small window or a filter in order to learn. The image is divided into small subparts, and multiple filters are used to study different local features such as edges and textures. In this way, these basic features are combined to create higher-order features - such as an eye and a hand, among others.

These filters are metrics of numbers that can be trained or updated. A CNN detects the necessary features from the image data with the correct trained filters and extracts the relevant predictive features (Figure 13.11). Filters are optimized with each iteration by using the mentioned backpropagation method. By design, convolution is a mathematical function on two objects which results in how one object is changed while operating with other the object. This process is also called feature mapping.

In addition, there are layers other than convolution layers; these are pooling and flattening. The method of pooling also slides a window on the pixels of an image and fuses local features into one feature. The pooling layers can be of many types:

Feature Extraction using convoluted layer

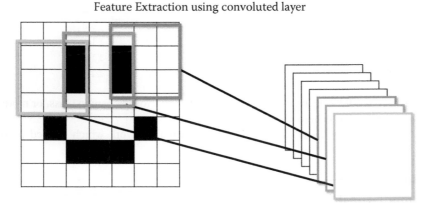

FIGURE 13.11 Feature Extraction Using CNNs.

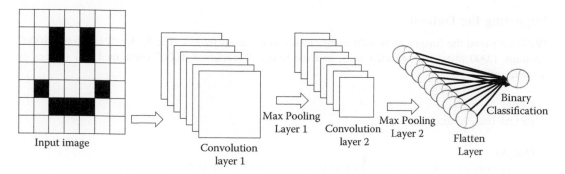

FIGURE 13.12 Architecture of a CNN Bases Neural System.

1. Max pooling is when the maximum intensity pixel is selected over the other pixels within the window.
2. Average pooling is when the average of all of the pixels is returned by a window.

Pooling is mainly intended to reduce the data size. At this step, the model retains important information among the local pixels such as the high intensity or the average of all of the pixels and drops other insignificant pieces of information.

After the extraction of local features from the images using CNNs and pooling operation, these features need to be fed to ANNs for classification. Therefore, these 2D features are converted into 1D features using the flattened layer, which is then inputted to the ANNs for further processing. Figure 13.12 shows a basic architecture of CNN based on artificial neural networks.

Implementation of CNN Using TensorFlow

In this project, we will use CNN to classify malaria parasite-infected cells and normal uninfected cells using their light microscopy images. The dataset is archived at NLM (IRB#12972) and was used in Rajaraman et al. (2018). The dataset contains 27,558 cell images with equal instances of parasitized and uninfected cells. However, to make this project computationally less exhaustive, we selected 1,000 images of each class, and these have been provided with the supplementary files of this chapter.

Import Libraries

For loading image data, we require certain additional libraries, as image data is generally comprised of individual files of images, and these have to be loaded one by one. Images belonging to each class are kept separately in different folders wherein we have to visit and read individual files. Therefore, we will require a built-in Python library called "os" which helps in handling files and directories. Additionally, we will require a library for handling image files, and the "Image" package will serve the purpose.

Code:
```
#Import some necessary Modules
import os
import numpy as np
from PIL import Image
import matplotlib.pyplot as plt
```

Importing the Dataset

We have saved the images of two different classes in different folders. The folder "Parasitized_1000" contains 1,000 images of infected cells, and the folder "Uninfected_1000" contains 1,000 images of normal cells.

Code:
```
print(os.listdir("./"))
```

Output:
```
['.ipynb_checkpoints', 'Parasitized_1000', 'Uninfected_1000']
```

The folders and files in a directory can be produced using the statement "listdir()" method of the OS module. It takes the path of the directory as an input parameter. To see the contents of the present directory, we can pass this string to the "listdir" method "./".

Code:
```
def image_loader(dr):
    IMG_list = []
    read = lambda im: np.asarray(Image.open(im).
                              convert("RGB").resize((64, 64)))
    for image_file in os.listdir(dr):
        Path = os.path.join(dr,image_file)
        _, ftype = os.path.splitext(Path)
        if ftype == ".png":
            img = read(Path)
            IMG_list.append(np.array(img)/255.)
    return IMG_list
Parasitized = np.array(image_loader('./Parasitized_1000'))
Uninfected = np.array(image_loader('./Uninfected_1000'))
Parasitized.shape
```

Output:
```
(1000, 64, 64, 3)
```

In the block of code above, we have mentioned a function named "Dataset_loader()". It uses the path of the directory where images are saved, reads each image, resizes them into 64 × 64 pixels, converts them, and appends them into a NumPy array. Lastly, it returns the array appended with all of the images. In the output, we can observe the shape of the "Parasitized" variable where 1,000 images of 64 × 64 pixels of three color channels (i.e. "RGB") are stacked together. We have resized the image to utilize less memory so that the example can be executed in any computer, or else holding image data will become a rather daunting and memory-intensive task.

Code:
```
Uninfected_label = np.zeros(len(Uninfected))
Parasitized_label = np.ones(len(Parasitized))
```

Here we have generated 1,000 zeros and 1,000 ones for labeling the dataset, where uninfected cells will be labeled as "0" and infected or parasitized cells will be labeled as "1".

Splitting the Dataset into the Training Set and Test Set

Code:

```
# Merge data
X = np.concatenate((Uninfected, Parasitized), axis = 0)
y = np.concatenate((Uninfected_label, Parasitized_label), axis = 0)

from sklearn.model_selection import train_test_split
X_train, X_test, y_train, y_test = train_test_split(X,y,test_size=0.3)
```

The images are assigned to the independent variable (i.e. "X"), and the labels are stored in variable "*y*". Next, the dataset is split randomly into training and testing, features, and labels, respectively. Let us take a look at the images in the training set. In the code block below, we create a canvas of size 15×15 inches when we are extracting 12 images and their labels from the training set and plot them on the canvas (Figure 13.13).

Code:

```
fig=plt.figure(figsize=(10,10))

for i in range(1, 13):
    ax = fig.add_subplot(3, 4, i)
    if y_train[i] == 0:
```

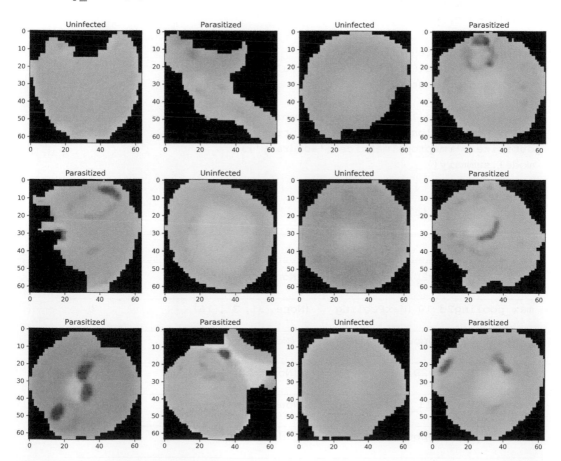

FIGURE 13.13 Sample Images from the Training Set.

```
        ax.title.set_text('Uninfected')
else:
        ax.title.set_text('Parasitized')
    plt.imshow(X_train[i], interpolation='nearest')
plt.show()
```

Output:

Building Model

Code:
```
from tensorflow.python.keras.models import Sequential
from tensorflow.python.keras.layers import Dense, Flatten, Conv2D, MaxPool2D,
Dropout

model = Sequential()
model.add(Conv2D(16,(3,3),activation='relu',input_shape=(64,64,3)))
model.add(MaxPool2D(2,2))
model.add(Dropout(0.2))

model.add(Conv2D(64,(3,3),activation='relu'))
model.add(MaxPool2D(2,2))
model.add(Dropout(0.5))

model.add(Flatten())
model.add(Dense(64,activation='relu'))
model.add(Dropout(0.5))

model.add(Dense(1,activation='sigmoid'))
model.compile(loss='binary_crossentropy',
optimizer='rmsprop',metrics=['accuracy'])
model.summary()
```

Output:
```
Model: "sequential_5"
```

Layer (type)	Output Shape	Param #
conv2d_10 (Conv2D)	(None, 62, 62, 16)	448
max_pooling2d_10 (MaxPooling	(None, 31, 31, 16)	0
dropout_15 (Dropout)	(None, 31, 31, 16)	0
conv2d_11 (Conv2D)	(None, 29, 29, 64)	9280
max_pooling2d_11 (MaxPooling	(None, 14, 14, 64)	0
dropout_16 (Dropout)	(None, 14, 14, 64)	0
flatten_5 (Flatten)	(None, 12544)	0

dense_10 (Dense)	(None, 64)	802880
dropout_17 (Dropout)	(None, 64)	0
dense_11 (Dense)	(None, 1)	65

```
=================================================================
Total params: 812,673
Trainable params: 812,673
Non-trainable params: 0
```

The model contains seven layers other than the dropout layers, and these are the following:

1. The convoluted layer with dimensions 3 × 3 and 16 filters with the ReLU activation function,
2. The max pool layer with size 2 × 2,
3. The second convoluted layer with 64 filters of 3 × 3 dimensions and the ReLU activation function,
4. The second max poll layer of size 2 × 2,
5. The flattened layer,
6. The fully connected dense layer with 64 neurons with the ReLU activation function, and
7. A single neuron with a sigmoid function for binary classification.

The model is compiled with the "binary_entropy" loss function and the "rmsprop" optimizer.

Traning Model

In this case, since the full epoch is too large, it will be difficult to train all of the epochs at a single time due to the limitation of memory. Therefore, we will divide the epoch into batches, then we will train one batch at a time. Here, we selected eight images per batch. For the rest, we added the early stopping callback (Figure 13.14).

Code:

```
from tensorflow.keras.callbacks import EarlyStopping
early_stop = EarlyStopping(monitor='val_loss', mode='min', verbose=1,
patience=5)
model.fit(x=X_train,
          y=y_train,
          epochs=400,
          batch_size= 8,
          validation_data=(X_test, y_test), verbose=1,
          callbacks=[early_stop]
          )
```

Output:

```
Train on 1400 samples, validate on 600 samples
Epoch 1/400
1400/1400 [==============================] - 6s 4ms/sample - loss:
0.7617 - accuracy: 0.5771 - val_loss: 0.6414 - val_accuracy: 0.7083
.
.
.
```

```
.
Epoch 34/400
1400/1400 [==============================] - 3s 2ms/sample - loss:
0.4088 - accuracy: 0.8700 - val_loss: 0.4210 - val_accuracy: 0.8217
Epoch 00034: early stopping

<tensorflow.python.keras.callbacks.History at 0x180cf150048>
```

Code:
```
import pandas as pd
model_loss = pd.DataFrame(model.history.history)
ax =model_loss[['loss','val_loss']].plot()
ax.set_xlabel('Epoch')
```

Output:
The plot in Figure 13.14 presents that the model is not overfitted.

Model Evaluation

Code:
```
predictions = model.predict_classes(X_test)
from sklearn.metrics import classification_report,confusion_matrix
print(confusion_matrix(y_test,predictions))
print(classification_report(y_test,predictions))
```
Output:
```
[[226  76]
 [ 31 267]]
```

	precision	recall	f1-score	support
0.0	0.88	0.75	0.81	302
1.0	0.78	0.90	0.83	298
accuracy			0.82	600
macro avg	0.83	0.82	0.82	600
weighted avg	0.83	0.82	0.82	600

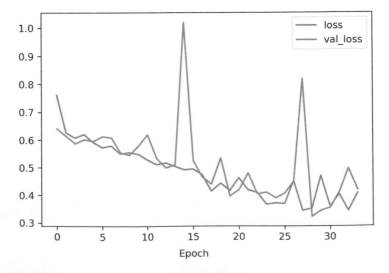

FIGURE 13.14 Plot of Training Loss and Validation Loss per Epoch.

The model has achieved a good accuracy rate of 82% on the test set. The model can be further improved by increasing the patience of the early stopping object, as we have learned that the model may uncertain spikes while training.

Exercise

1. How is the output calculated using forward propagation in neural networks?
2. How can gradient decent help in updating the weights in neural networks?
3. Explain the steps of backpropagation during the training of ANNs.
4. Why are neural networks prone to overfitting, and how can we evaluate if an ANN is overfitted?
5. What are the techniques used to overcome the overfitting problem in neural networks?
6. Apply ANN on the diabetes dataset from chapter 11.

REFERENCE

Rajaraman, S., Antani, S. K., Poostchi, M., Silamut, K., Hossain, M. A., Maude, R. J., Jaeger, S., & Thoma, G. R. (2018). Pre-trained convolutional neural networks as feature extractors toward improved malaria parasite detection in thin blood smear images. *PeerJ*. https://doi.org/10.7717/peerj.4568.

14

The Machine Learning Project

Introduction

Up until this point, we have learned five supervised learning algorithms for classification - specifically, logistic regression, *k*-nearest neighbors, random forests, support vector machines, and artificial neural networks. In this chapter, we will go through a real-world problem from start to end- from data retrieval to selecting a model based on their performance on the dataset. We have already learned that various classification algorithms have advantages and disadvantages - for example, logistic regression and SVMs perform well for linearly separable data, whereas *K*-NN, decision trees, and ANNs can have nonlinear classification boundaries.

The first step of a Machine Learning project is the retrieval of data and the selection of appropriate features. In this example, we will use sequences of proteins encoded by genes to train models to discover their association with age-related disorders.

The quality of life during old age is typically much lower than in youth, since elderly people often suffer from multiple age-related disorders (ARDs) - such as dementia, arteriosclerosis, osteoporosis, diabetes, osteoarthritis, or cancer. The cause of death may also be linked to gradual atrophies of the tissue, neuropathy, or microvascular leakage. The prediction of disease genes is typically a classification issue in which the genes that are associated with diseases are differentiated from benign genes. Techniques in Machine Learning can be used to classify a similar set of genes based on their features with other sets. In this example, we will try to train models that can identify genes associated with age-related diseases. The dataset and features used here are a subset of the dataset used by (Srivastava et al., 2016). The subset of features includes the sequence-based properties of proteins that are encoded by the genes - such as hydrophobicity, polarity, polarizability, secondary structure, Van der Waals volume, charge, solvent accessibility, etc. These features can be retrieved from the protein sequences using tools like PROFEAT (https://www.hsls.pitt.edu/obrc/index.php?page=URL1153154153). While there a number of other features like network interaction features and cellular location features, we will use only the sequence-based features to make the dataset small enough to be trained in any local machine. The objective of this chapter is to test the supervised algorithms that we have learned so far from Chapters 9–13 to evaluate their performance on a single dataset. Furthermore, the Jupyter Notebook provided in the supplementary files of this chapter can be used as a template to implement several algorithms all at once.

Importing the Libraries

We will start by importing the required libraries.
Code:
```
import pandas as pd
import seaborn as sns
import matplotlib.pyplot as plt
import numpy as np
```

Importing the Dataset

The dataset is saved in Excel formal and is named as "genes_data.xlsx". It can be found in the supplementary file with this chapter.

Code:
```
dataset = pd.read_excel('protein_dataset.xlsx')
dataset.shape
```
Output:
```
(658, 75)
```
The dataset has 659 instances and 75 features. All of these can be calculated with protein sequences. Some of the important features of this are:

1. Hydrophobicity
2. Normalized Van der Waals
3. Polarity
4. Polarizability
5. Charge
6. Solvent Accessibility
7. Secondary Structure
8. Surface Tension
9. Protein-Protein Interface Hotspot Propensity
10. Protein-DNA Interface Propensity
11. Protein-RNA Interface Propensity
12. Molecular Weight
13. Solubility in Water
14. Amino Acid Flexibility Index

Most of the features have subgroups within themselves, making the total number features into 74, and the last column are the labels wherein age-related genes are labeled as "1" and benign genes are labeled as "0".

Code:
```
dataset['Target'].value_counts()
```
Output:
```
1   329
0   329
Name: Target, dtype: int64
```
The dataset has equal instances of positive and negative sets, so the base accuracy rate is 50%. Next, we separate the dependent and independent variables (i.e. "*y*" and "*x*"), respectively.

Code:
```
X = dataset.iloc[:,2:-1].values
y = dataset.iloc[:,-1].values
```
The features are scaled using sklearns's "StandardScaler" preprocessing class.

Code:
```
from sklearn.preprocessing import StandardScaler
sc = StandardScaler()
X_norm = sc.fit_transform(X)
```

PCA

Let us try to use the PCA to visualize the distribution instances in two-dimensional space.

Code:
```
from sklearn.decomposition import PCA
sklearn_pca = PCA(n_components=2)
PCs = sklearn_pca.fit_transform(X_norm)
data_transform = pd.DataFrame(PCs,columns=['PC1','PC2'])
data_transform = pd.concat([data_transform,dataset.iloc[:,-1]],axis=1)
```

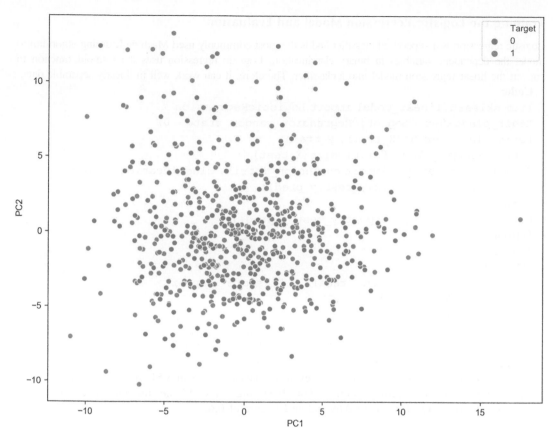

FIGURE 14.1 PCA Plot for the Dataset.

```
fig, axes = plt.subplots(figsize=(10,8))
sns.set_style("whitegrid")
sns.scatterplot(x='PC1',y='PC2',data = data_transform,hue='Target',s=60,
cmap='gre
  y')
```

Output:

From Figure 14.1, it is important to note that the instances are well blended based on the sequence features in two-dimensional space.

Splitting the Dataset into the Training Set and the Test Set

Splitting the dataset is important in evaluating the models in the test set, as we have discussed in the previous chapters. In this case, sklearn's "train_test_split" processer will be used for the purpose.

Code:

```
from sklearn.model_selection import train_test_split
X_train, X_test, y_train, y_test = train_test_split(X_norm, y, test_size =
0.20, rand
  om_state = 101)
```

Now, we will apply the training set to all of the five classification algorithms that we have learned so far and assess each of them using the test set. We will start with logistic regression.

Training the Logistic Regression Model and Evaluation

Logistic regression is a supervised classifier and is the most commonly used Machine Learning algorithm to model the dependent variables in binary classification. Logistic regression uses the sigmoid function to convert the linear regression model into a classifier. Therefore, it can work well in linearly separable data.

Code:

```
from sklearn.linear_model import LogisticRegression
Logit_classifier = LogisticRegression(random_state = 0)
Logit_classifier.fit(X_train, y_train)
y_pred = Logit_classifier.predict(X_test)
from sklearn.metrics import confusion_matrix, classification_report
print(confusion_matrix(y_test,y_pred))
print('\n')
print(classification_report(y_test,y_pred))
```

Output:

```
[[14 29]
 [24 53]]
```

	precision	recall	f1-score	support
0	0.37	0.33	0.35	43
1	0.65	0.69	0.67	77
accuracy			0.56	120
macro avg	0.51	0.51	0.51	120
weighted avg	0.55	0.56	0.55	120

While training the logistic regression model in the dataset, we achieved an accuracy rate of 56%, which is more than the baseline accuracy. For classifying class "1" (i.e. the age-related genes) the algorithm achieved a recall value of 0.69 and an F1 score of 0.6.

Training the *K*-NN Model and Evaluation

K-nearest algorithm is one of the simplest supervised Machine Learning models where the predictions are made based on the nearest datapoints of the instance, so, this can be used for classifying linearly non-separable datasets. Selecting the value of "*k*" is rather crucial for the performance of this algorithm. Therefore, we will train the algorithm with various "*K*" values and then determine the *k*-value with minimum error.

Choosing K-Value

Code:

```
from sklearn.neighbors import KNeighborsClassifier
errors = []
for i in range(1,100):
    knn = KNeighborsClassifier(n_neighbors=i)
    knn.fit(X_train,y_train)
    pred_i = knn.predict(X_test)
    errors.append(np.mean(pred_i != y_test))
plt.figure(figsize=(10,6))
plt.plot(range(1,100),errors,color='black', linestyle='dashed',
marker='o',
        markerfacecolor='black', markersize=10)
plt.title('Error Rate vs. K Value')
plt.xlabel('K Value')
plt.ylabel('Error Rate')
```

Output:

```
Text(0, 0.5, 'Error Rate')
```
Based on Figure 14.2, we can observe that the error is lowest at "$k = 7$", so let us evaluate the model based on seven nearest neighbors.

Code:

```
from sklearn.neighbors import KNeighborsClassifier
KNN_classifier = KNeighborsClassifier(n_neighbors = 7)
KNN_classifier.fit(X_train, y_train)
y_pred = KNN_classifier.predict(X_test)
from sklearn.metrics import confusion_matrix, classification_report
print(confusion_matrix(y_test,y_pred))
print('\n')
print(classification_report(y_test,y_pred))
```

Output:

```
[[23 20]
 [24 53]]
```

	precision	recall	f1-score	support
0	0.49	0.53	0.51	43
1	0.73	0.69	0.71	77
accuracy			0.63	120
macro avg	0.61	0.61	0.61	120
weighted avg	0.64	0.63	0.64	120

The *K*-NN model attained an accuracy rate of 63%, which is more than the logistic regression model. Moreover, the precision for identifying age-related genes is 0.73 with an F1 score of 0.71.

Training the Random Forest Model and Evaluation

The random forest model is constructed from a combination of various decision trees trained on randomly selected data samples. The random forest model is also known as an ensemble method. The

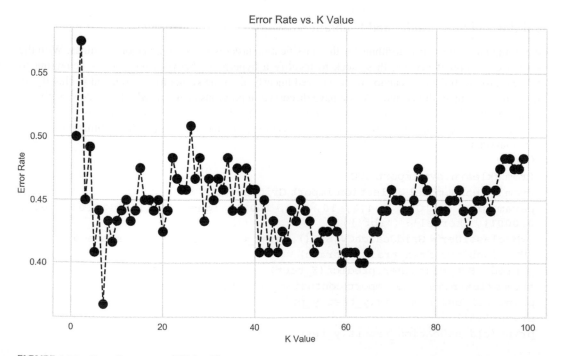

FIGURE 14.2 Error Rate versus *K*-Value Chart.

performance of the model depends upon the number of decision trees generated on a random subset of data. Therefore, we will use the grid search technique with a list of "n_estimator" values for optimizing the number of trees in the random forest model.

Code:

```
from sklearn.model_selection import GridSearchCV
from sklearn.ensemble import RandomForestClassifier
param_grid = {'n_estimators': [10, 100,150,200,250,300,350,400]}
RF_classifier = GridSearchCV(RandomForestClassifier(),param_grid,refit=True,
verbose=
0)
RF_classifier.fit(X_train,y_train)
y_pred = RF_classifier.predict(X_test)
from sklearn.metrics import confusion_matrix, classification_report
print(confusion_matrix(y_test,y_pred))
print('\n')
print(classification_report(y_test,y_pred))
```

Output:

```
[[21 22]
 [23 54]]
```

	precision	recall	f1-score	support
0	0.48	0.49	0.48	43
1	0.71	0.70	0.71	77
accuracy			0.62	120
macro avg	0.59	0.59	0.59	120
weighted avg	0.63	0.62	0.63	120

The random forest model achieved an accuracy rate of 62%, which is slightly less than the K-NN model, while the dataset has a baseline accuracy rate of 50%. The model has precision and F1 scores of 0.71 for identifying the class "1" or the age-related genes.

Training the SVM Model and Evaluation

One of the most common algorithms for the classification tasks is supporting vector machines. With the assistance of support vectors, these seek to identify a hyperplane between data points belonging to different groups. If datasets cannot be categorized linearly, then the kernel trick is utilized in classifying the features to higher dimensions. As we have discussed in previous chapters, SVM is quite sensitive to its parameters like "c", the soft margin, "gamma", and the effect of generated features on the model. Therefore, it is apparent here as well that we will use the grid search technique for optimization of the SVM classifier.

Code:

```
from sklearn.svm import SVC
from sklearn.model_selection import GridSearchCV
param_grid = {'C': [0.1,1, 10, 100, 1000], 'gamma': [1,0.1,0.01,0.001,
0.0001], 'kernel': ['rbf']}
SVM_classifier = GridSearchCV(SVC(),param_grid,refit=True,verbose=0)
SVM_classifier.fit(X_train,y_train)
y_pred = SVM_classifier.predict(X_test)
from sklearn.metrics import confusion_matrix, classification_report
print(confusion_matrix(y_test,y_pred))
print('\n')
print(classification_report(y_test,y_pred))
```

Output:
```
[[10 33]
 [20 57]]
              precision    recall    f1-score    support
         0       0.33       0.23       0.27         43
         1       0.63       0.74       0.68         77
  accuracy                             0.56        120
 macro avg       0.48       0.49       0.48        120
weighted avg     0.53       0.56       0.54        120
```
SVM shows lower accuracy than the *K*-NN and random forest model, which is 56%. While for predicting the positive class, it has a recall of 0.74.

Training the ANN Model and Evaluation

Artificial neural networks (ANNs), another commonly used supervised learning algorithm, is based on the functioning of biological neurons. A neuron produces output by calculating a weighted sum of input characteristics, which then proceeds through an activation function. Neurons learn by updating their weights. The weights are modified by means of an algorithm called gradient descent. The gradient descent algorithm utilizes derivatives to backpropagate errors and updates the weights in a way that enhances network efficiency.

Code:
```
import tensorflow as tf
from tensorflow.keras.models import Sequential
from tensorflow.keras.layers import Dense, Activation, Dropout
from tensorflow.keras.callbacks import EarlyStopping
model = Sequential()
#
model = Sequential()
model.add(Dense(units=60,activation='relu'))
model.add(Dropout(0.5))
model.add(Dense(units=15,activation='relu'))
model.add(Dropout(0.5))
model.add(Dense(units=1,activation='sigmoid'))
model.compile(loss='binary_crossentropy', optimizer='rmsprop',metrics=
['accuracy'])
early_stop = EarlyStopping(monitor='val_loss', mode='min', verbose=1,
patience=5)model.fit(x=X_train,
y=y_train,
epochs=400,
batch_size= 64,
validation_data=(X_test, y_test), verbose=1,
callbacks=[early_stop])
```
Output:
```
Train on 478 samples, validate on 120 samples
Epoch 1/400
478/478 [==============================] - 2s 5ms/sample - loss: 0.8473 -
accuracy: 0.5335 - val_loss: 0.6755 - val_accuracy: 0.6250
Epoch 2/400
478/478 [==============================] - 0s 278us/sample - loss: 0.8412 -
accuracy: 0.5167 - val_loss: 0.6789 - val_accuracy: 0.5750
```

```
Epoch 3/400
478/478 [==============================] - 0s 275us/sample - loss: 0.7813 -
accuracy: 0.5188 - val_loss: 0.6821 - val_accuracy: 0.5500
Epoch 4/400
478/478 [==============================] - 0s 290us/sample - loss: 0.7877 -
accuracy: 0.5272 - val_loss: 0.6777 - val_accuracy: 0.5917
Epoch 5/400
478/478 [==============================] - 0s 293us/sample - loss: 0.7411 -
accuracy: 0.5397 - val_loss: 0.6802 - val_accuracy: 0.5750
Epoch 6/400
478/478 [==============================] - 0s 283us/sample - loss: 0.7658 -
accuracy: 0.5230 - val_loss: 0.6861 - val_accuracy: 0.5583
Epoch 00006: early stopping
```

This neural network has two densely connected layers with 60 and 15 neurons in each layer, respectively. The last layer or the output layer has one neuron with a sigmoid activation function. Dropout layers were added in between each layer to minimize the risk of overfitting. The model is compiled with "binary_crossentropyloss" and "rmsprop" optimizer. Lastly, it was trained with an early stopping callback function to disrupt the training loop once training loss is less than validation loss. The model stopped training after six epochs. Next, we evaluated the model on the test set to generate the classification report.

Code:

```
predictions = model.predict_classes(X_test)
from sklearn.metrics import classification_report,confusion_matrix
print(confusion_matrix(y_test,predictions))
print(classification_report(y_test,predictions))
```

Output:

```
[[14 29]
 [24 53]]
              precision    recall  f1-score   support
           0       0.37      0.33      0.35        43
           1       0.65      0.69      0.67        77
    accuracy                           0.56       120
   macro avg       0.51      0.51      0.51       120
weighted avg       0.55      0.56      0.55       120
```

The ANN model has an accuracy rate of 56%.

Predicting genes associated with a group of diseases such as age-related diseases is primarily a special type of Machine Learning problem called positive unlabeled learning (PU learning) where we do not carry a well defined true negative set. In PU learning, the positive set is well-defined or known, but generating a negative set is based on the assumption that the instances that do not belong to the positive set can be treated as part of the a negative set. The performance of the model is measured by evaluating it for predicting a positive set. In this way, the K-NN and random forest models' performance were better in comparison with the rest of the models. This can be due to the fact that these models are capable of classifying the nonlinearly separable dataset, and we estimate an idea of the nonlinearity of the data from the PCA plot (Figure 14.1). Furthermore, the overall accuracy can be increased by adding various network and cell location-based features, as shown in (Srivastava et al., 2016), because sequence-based parameters do not completely define the function of genes. Readers can efficiently use the Jupyter Notebook of this chapter for an easy-to-use template for the implementation of the five classification algorithms on most of the datasets.

Exercise

1. Apply all of the datasets provided for training various algorithms in the supplementary files of the previous respective chapters to this chapter's template and compare the models with one another.
2. What is positive unlabeled learning?
3. Why is the classification of genes associated with a disease that mostly falls in the PU learning category? Discuss.

REFERENCE

Srivastava, I., Gahlot, L. K., Khurana, P., & Hasija, Y. (2016). DbAARD & AGP: A computational pipeline for the prediction of genes associated with age related disorders. *Journal of Biomedical Informatics*. https://doi.org/10.1016/j.jbi.2016.01.004

15

Natural Language Processing

Introduction

Until this point, we have seen data like numerical values and images as input to Machine Learning algorithms. In this chapter, we will use text data as input to Machine Learning. Natural language processing, or NLP, is a subfield of Machine Learning where we process text data and use Machine Learning algorithms to perform various tasks like classification, generation of new text, identification of associated words, automatic summarization of articles, translation of languages, etc. Applications of NLP can be found everywhere – from social media platforms, search engines, chatbots, translators, and many more. In the domain of biology, there is an ocean of information buried in the form of research articles. In performing tasks like data curation, finding entity associations like drug-protein interaction, protein-protein interactions, gene-disease association, genotype-phenotype association, among others, curators have to go through all of the text literature for creating databases. Nowadays, these curators use NLP techniques to classify the texts at the first step to minimize the number of articles and then use other sophisticated NLP algorithms to find the associations as mentioned above. NLP is also used to summarize long articles while preserving important sentences. Another promising application of NLP is in processing genomic or protein sequences, because this is also in text formats. NLP can classify or cluster sequences, predict new instances from raw sequences and has many more useful applications.

Text data is unstructured data – the vocabulary of which is significantly large. Different words can have the same meaning, and even the same words can have different definitions, unlike programming languages such as Python which are very well structured. We have studied in the previous chapter that images have to be converted into arrays based on pixel intensities in order to be used in Machine Learning algorithms. In the same way, to input text data to a Machine Learning algorithm, the text also needs to be vectorized, more specifically, it needs to be converted into numeric representations with features before being applied in any Machine Learning algorithm.

Generally, there are three steps for any NLP task:

1. Data aggregation is where we accumulate or compile text data (i.e. articles, abstracts, gene sequences, or protein sequences, etc) that are relevant to the objective of the task.
2. The text data must be represented in a vector format.
3. A Machine Learning algorithm is used to analyze it.

To reiterate, the collection of data is the first step for any Machine Learning task. While compiling data, the quality of the data should be maintained. For more information, please refer to chapter 8 – "Challenges with Data" subsection. The collected text data is called the text corpus.

Vectorizing the Text

Bag of Words

Supposing we have two sentences in our text corpus:

1. "Metformin is used for treating diabetes."
2. "Amlodipine is used for treating hypertension."

Table 15.1 is the vector representation of the two sentences. The columns represent the vocabulary of the corpus, and the rows serve as the sentences. The sentence will be encoded based on the presence or absence of words in the vocabulary. In Table 15.1, the first row is an encoded version of the first sentence. We can observe that, in the first statement, the values in the columns for "Amlodipine" and "hypertension" are "0", because both do not occur in the statement, and the values in the rest of the columns are "1", which is the number of times these words appear in the sentence. In the same way, the vectorized representation of the second sentence is presented. If in case a word repeats twice in the text, then its vector representation will be "2". This type of representation of text is known as "one-hot encoding" or "bag-of-words". These features generally have high dimension, because vocabulary size can exceed by more than thousands and is maximally filled with zeros called sparse matrixes. Since the text is converted into features, we can apply Machine Learning algorithms to these. We can even perform simple mathematical operations, such as finding their distance in space using Euclidean distance.

TABLE 15.1

Vector Representation of Text

"Metformin"	"Amlodipine"	"is"	"used"	"for"	"treating"	"diabetes"	"hypertension"
1	0	1	1	1	1	1	0
0	1	1	1	1	1	0	1

While "bag-of-words" can convert the text into vectors, it generalizes all of the words as the same; in particular, it does not have any weightage for the word itself based on, for example, the importance of the word in a sentence or the uniqueness of a word in the corpus.

TF-IDF

We can assign importance to the words and their uniqueness based on the word's frequency in the corpus. Therefore, we calculate "term frequency–inverse document frequency", or the TF-IDF for the words. In this term, TF is defined as the importance of the word within the text instance. A text instance can be a sentence, a paragraph, or an article. Term frequency is calculated by counting the occurrence of a word in the text instance. The IDF or the inverse document frequency is the importance of the word in the whole corpus. IDF is calculated with the following formula:

$$\text{IDF}_{(word)} = \log(D/I_{(word)})$$

where $\text{IDF}_{(word)}$ is the importance of a word in the document; D is the total number of instances in the corpus; $I_{(word)}$ is the number of text instances of the particular word.

Simply put, IDF is the inverse of the number of occurrences of a word in a corpus. We calculate the inverse because it is our goal to give the unique words more importance, since certain words, such as auxiliary verbs (i.e. "be", "do", and "have") and articles (i.e. "a", "an", and "the") occur often but do not add significant information. On the other hand, words like the name of drugs, proteins, or diseases will appear less but are much more informative. Therefore, we inverse the frequency of the words.

The value for a word in a text instance is the product of TF and IDF. Furthermore, the importance and uniqueness of the words will be preserved.

Next, now we have acquired vector representations for the text data. Let us implement some NLP tasks using Python.

Classification of Abstracts into Various Categories Using NLP

In this section, we will use sample data from (Hersh et al., 1994). The dataset is a collection of titles and abstracts grouped into 19 classes of diseases. We have selected four groups out of these – specifically, "Bacterial Infections and Mycoses", "Skin and Connective Tissue Diseases", "Virus Diseases", and "Eye Diseases". The goal of this task is to build a classifier that can categorize any new given abstract into these four categories. The dataset is provided as a Microsoft Excel file and is named "NLP_data" in the supplementary file with this chapter. We will use a bag-of-words and the TF-DIF for vectorization of the text data and build a classifier using logistic regression. Lastly, we will present a deep learning implementation of the same data.

Importing the Dataset

The Pandas library has a method called "read_excel" to read .xlsx files (Table 15.2).

TABLE 15.2

Sample of Text Data

	Abstract	Label
0	Activity of temafloxacin against respiratory p...	Bacterial_Infections_and_Mycoses
1	Corneal ulcer caused by Pseudomonas pseudomall...	Bacterial_Infections_and_Mycoses
2	Induction of suppressor T cells and inhibition...	Skin_and_Connective_Tissue_Diseases
3	Xanthoma in Meckel's cave. A case report. A ca...	Skin_and_Connective_Tissue_Diseases
4	Analysis of IL-1 and TNF-alpha gene expression...	Skin_and_Connective_Tissue_Diseases

Code:
```
dataset = pd.read_excel('NLP_data.xlsx')
dataset.head()
```

Output:
In this example, we have two columns. The first one contains the abstracts, and the second contains the labels. In this case, labels are the dependent variables.

Code:
```
dataset['Label'].value_counts()
```

Output:
```
Bacterial_Infections_and_Mycoses        2198
Skin_and_Connective_Tissue_Diseases     1457
Virus_Diseases                           925
Eye_Diseases                             842
Name: Label, dtype: int64
```
The "value counts" method on the "Label" column produces the number of instances for each class.

Text Processing

For text processing, we will require an additional library called "nltk" or natural language tool kit. This library contains various methods for dealing with text data. In this part, we utilize the "nltk" library for removing the stopwords such as "the", "I", "we", etc., which are considered as noise, because they have a high frequency in any text data but have a minimum contribution to the information about a text. The "nltk" library can be installed using "pip" or "conda" installer.

Code:

```
import nltk
nltk.download('stopwords')
```

Once the library is installed, we will download the list of stopwords using the download method. The "nltk" also contains various corpora as example datasets to practice NLP tasks. Let us check the contents of "stopwords".

Code:

```
from nltk.corpus import stopwords
stopwords.words('english')[0:15]
```

Output:

```
['i',
 'me',
 'my',
 'myself',
 'we',
 'our',
 'ours',
 'ourselves',
 'you',
 "you're",
 "you've",
 "you'll",
 "you'd",
 'your',
 'yours']
```

In the output, we are presented with the top 15 stopwords from the list. Next, we will write a function to remove these stopwords from the text of the abstracts.

Code:

```
def text_processing(text):
    word_list = text.lower().split(" ")
    processed_words = []
    for word in word_list:
        if word not in stopwords.words('english'):
            processed_words.append(word)
    return " ".join(processed_words)
```

This is a simple function where we take the text as input, then we split the text into a list of words. These words are examined for the presence of words in the "stopword" corpus. The words that are not found in the corpus are retained and joined to form a text.

Code:
```
text = 'Hi we are discussing NLP in this chapter'
text_processing(text)
```

Output:
```
'hi discussing nlp chapter'
```

In the output, we can observe that the function has removed "we", "are", "in", and "this" from the original text, and the new text to be parsed now contains more informative keywords. In the next step, we will apply this function to the "Abstract" column of the dataset.

Code:
```
X = dataset['Abstract'].apply(text_processing)
dataset['Abstract'] = X
```

As the text of abstracts is now exclusive of stopwords, we will separate dependent and independent variables.

Code:
```
X = dataset['Abstract']
y = dataset['Label']
```

Label Encoding

Code:
```
from sklearn.preprocessing import LabelEncoder
le = LabelEncoder()
y = le.fit_transform(y)
y[:5]
```

Output:
```
array([0, 0, 2, 2, 2])
```

To convert our labels into numeric representations for each class, we used the "LabelEncoder" class of sklearn. As we can see, the first "five" items in the output have integers for classes.

Text Tokenization Bag-of-Words

In this section, we will implement text vectorization. We will convert the sentences into the bag-of-words model. Sklearn has certain important feature extraction libraries for text data. The class "CountVectorizer" will convert the abstracts into the bag-of-words model, where each column represents the words, and the column entries stand for the word counts for these previously mentioned words in the abstract.

Code:
```
from sklearn.feature_extraction.text import CountVectorizer
cv = CountVectorizer()
X = cv.fit_transform(X).toarray()
X.shape
```

Output:

```
(5422, 25953)
```

After transforming "*X*", we can observe that the data matrix is of the size 5,422 × 25,953, implying that the corpus has 25,953 words in its vocabulary, and we have 5,422 instances or text. "25,953" is a significantly large dimension. Let us apply PCA on it to reduce the dimension to "2" and to visualize the text instances in 2D space.

Code:

```
from sklearn.decomposition import PCA
sklearn_pca = PCA(n_components=2)
PCs = sklearn_pca.fit_transform(X)
PCA_x = pd.DataFrame(PCs,columns=['PC1','PC2'])
PCA_x = pd.concat([PCA_x,dataset['Label']],axis=1)
fig, axes = plt.subplots(figsize=(10,8))
sns.set_style("whitegrid")
sns.scatterplot(x='PC1',y='PC2',data = PCA_x,hue='Label')
```

Output:

We can observe that the abstract of the virus disease class is more scattered, and insufficient information can be retrieved from this PCA plot (Figure 15.1). Let us feed the vectorized data into the logistic regression algorithm and study how it performs with this data.

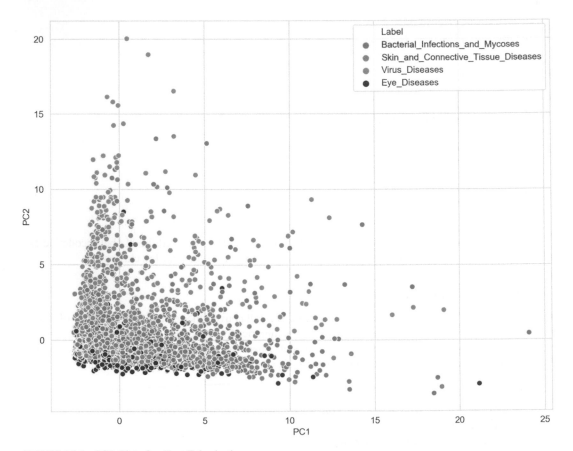

FIGURE 15.1   PCA Plot after Text Tokenization.

Splitting the Dataset into the Training Set and the Test Set

Code:
```
from sklearn.model_selection import train_test_split
X_train, X_test, y_train, y_test = train_test_split(X,
y,test_size=0.3,random_state=101)
```

Building Model

Code:
```
from sklearn.linear_model import LogisticRegression
classifier = LogisticRegression(random_state = 0)
classifier.fit(X_train, y_train)
```

Model Evaluation

Code:
```
predictions = classifier.predict(X_test)
from sklearn.metrics import confusion_matrix,classification_report
print(confusion_matrix(y_test,predictions))
print('\n')
print(classification_report(y_test,predictions))
```

Output:
```
[[597    7   25  14]
 [ 15  203   24   8]
 [ 26    2  406  10]
 [ 45    2   11 232]]
```

	precision	recall	f1-score	support
0	0.87	0.93	0.90	643
1	0.95	0.81	0.88	250
2	0.87	0.91	0.89	444
3	0.88	0.80	0.84	290
accuracy			0.88	1627
macro avg	0.89	0.86	0.88	1627
weighted avg	0.89	0.88	0.88	1627

The logistic regression was able to prioritize the features for respective classes and yielded an accuracy rate of 88%.

TF-IDF Implementation

In this section, we will implement TF-IDF using the "TFidFTransformer" class of sklearn. It collects the word count data (i.e. the transformed data using the "CountVectorizer" object) and applies the TF-IDF to every text instance.

Code:
```
from sklearn.feature_extraction.text import TfidfTransformer
tfidf_transformer = TfidfTransformer()
X = tfidf_transformer.fit_transform(X)
print(X.shape)
X = X.toarray()
```

Output:
```
(5422, 25953)
```

We can retrieve the IDF (i.e. inverse document frequencies) for words from the "cv.vocabulary_" dictionary. In this dictionary, words are the keys, and values represent the IDF scores.

Code:
```
print(tfidf_transformer.idf_[cv.vocabulary_['lungs']])
```

Output:
```
5.934842800711414
```

The output shows the IDF score for the word "lungs". Again, we plot PCA by transforming the TF-IDF features into two dimensions.

Code:
```
from sklearn.decomposition import PCA
sklearn_pca = PCA(n_components=2)
PCs = sklearn_pca.fit_transform(X)
PCA_x = pd.DataFrame(PCs,columns=['PC1','PC2'])
PCA_x = pd.concat([PCA_x,dataset['Label']],axis=1)
fig, axes = plt.subplots(figsize=(10,8))
sns.set_style("whitegrid")
sns.scatterplot(x='PC1',y='PC2',data = PCA_x,hue='Label')
```

Figure 15.2 presents the PCA plot of TF-IDF transformed data reduced to two dimensions. In this plot, we can see the similarity of instances within the classes, as well as the differences between the classes. Next, we will apply the TF-IDF transformed data to train logistic regression again.

Splitting the Dataset into the Training Set and the Test Set

Code:
```
from sklearn.model_selection import train_test_split
X_train, X_test, y_train, y_test = train_test_split(X, y,test_size=0.2)
```

Building Model

Code:
```
from sklearn.linear_model import LogisticRegression
classifier = LogisticRegression(random_state = 0)
classifier.fit(X_train, y_train)
```

Model Evaluation

Code:
```
predictions = classifier.predict(X_test)
from sklearn.metrics import confusion_matrix,classification_report
print(confusion_matrix(y_test,predictions))
print('\n')
print(classification_report(y_test,predictions))
```

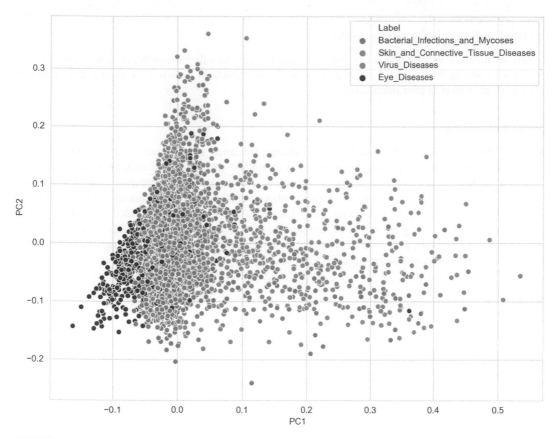

FIGURE 15.2 PCA Plot after TF-IDF Conversation.

Output:

```
[[440    0    9    9]
 [ 20  115   19    1]
 [ 37    0  248    2]
 [ 46    0    8  131]]
               precision  recall  f1-score  support

           0       0.81    0.96      0.88      458
           1       1.00    0.74      0.85      155
           2       0.87    0.86      0.87      287
           3       0.92    0.71      0.80      185

    accuracy                         0.86     1085
   macro avg       0.90    0.82      0.85     1085
weighted avg       0.87    0.86      0.86     1085
```

This time, we obtained an accuracy rate that is lower than the previous problem by 2%. This could be attributed to the fact that we are accounting for 5,422 text instances – which is far less than the number of features or vocabulary as comprated to 25,953 – for the calculation of TF-IDF scores. As the corpus increases, the TF-IDF model will perform better than the CountVectorizer transformations.

Artificial Neural Networks in NLP

In this section, we will implement neural networks for the classification problem. The application of neural networks in NLP is also known as deep NLP. As vectorized texts have a lot of features, neural networks are preferred over other algorithms.

Label Binarizing

To apply the concept of ANN to NLP, we will change the labels into the one-versus-all format. Until this point, we were handling binary classification in ANNs, where we had only two classes. In this problem, we have four classes, so we have to convert our labels into a multiclass classification case (Table 15.3). In the one-versus-all method, the number of class labels in the dataset and the number of neurons in the last layer have to be equal. The task of each neuron will be divided to predict for one class only. Simply put, only the respective assigned class will be positive for each neuron, and the rest will all be "0". Therefore, each neuron performs binary classification, but, in combination, they will be working to predict different classes (Table 5.3).

TABLE 15.3

Label Binarizing

Labels	Label Binarizing			
"Bacterial_Infections_and_- Mycoses"	1	0	0	0
"Eye_Diseases"	0	1	0	0
"Skin_and_Connective_Tissu- e_Diseases"	0	0	1	0
"Virus_Diseases"	0	0	0	1

Code:

Sklearn's "LabelBinarizer" transforms the labels into binary classes.

```
y = dataset['Label']
from sklearn.preprocessing import LabelBinarizer
lb = LabelBinarizer()
y = lb.fit_transform(y)
y.shape
```

Output:

```
(5422, 4)
```

The "LabelBinarizer" object's class_ attribute stores the array of all the original classes so that we can retrieve it in the future during prediction.

Code:

```
lb.classes_
```

Output:

```
array(['Bacterial_Infections_and_Mycoses', 'Eye_Diseases',
       'Skin_and_Connective_Tissue_Diseases', 'Virus_Diseases'],
       dtype='<U35')
```

Splitting the Dataset into the Training Set and the Test Set

Let us split the TF-IDF transform features and binarized labels into training and test sets.

Code:
```
from sklearn.model_selection import train_test_split
X_train, X_test, y_train, y_test = train_test_split(X, y,test_size=0.2,
random_state=101)
```

Building Model

We will build the ANN model using TensorFlow's Keras library, as discussed in the previous chapter.

Code:
```
import tensorflow as tf
from tensorflow.keras.models import Sequential
from tensorflow.keras.layers import Dense, Activation, Dropout
model = Sequential()
#
model.add(Dense(units=32,activation='relu'))
model.add(Dropout(0.5))
model.add(Dense(units=16,activation='relu'))
model.add(Dropout(0.5))
model.add(Dense(units=4,activation='softmax'))
# For a binary classification problem
model.compile(loss='categorical_crossentropy', optimizer='adam',metrics=
["accuracy"])
```

The model has three layers. The first layer contains 32 neurons that are densely connected to the second layer which has 16 neurons, and the last layer as four neurons because we are tasked to predict four classes. The last layer's activation function is "softmax", and we have applied the "adam" optimizer. There are dropouts of 0.5 present in between the layers. The model will be trained using the EarlyStopping callback object.

Training Model

Code:
```
from tensorflow.keras.callbacks import EarlyStopping
early_stop = EarlyStopping(monitor='val_loss', mode='min', verbose=1,
patience=1)
model.fit(x=X_train,
        y=y_train,
        epochs=400,
        batch_size=128,
        validation_data=(X_test, y_test), verbose=1,
        callbacks=[early_stop]
        )
```

Output:
```
Train on 4337 samples, validate on 1085 samples
Epoch 1/400
4337/4337 [==============================] - 7s 2ms/sample - loss: 1.0625 -
accuracy: 0.5635 - val_loss: 0.5607 - val_accuracy: 0.8682
```

```
Epoch 2/400
4337/4337 [==============================] - 4s 962us/sample - loss: 0.4298 -
accuracy: 0.8697 - val_loss: 0.3167 - val_accuracy: 0.8977
Epoch 3/400
4337/4337 [==============================] - 4s 968us/sample - loss: 0.1935 -
accuracy: 0.9477 - val_loss: 0.3038 - val_accuracy: 0.9041
Epoch 4/400
4337/4337 [==============================] - 4s 932us/sample - loss: 0.1084 -
accuracy: 0.9737 - val_loss: 0.3096 - val_accuracy: 0.9097
Epoch 00004: early stopping
<tensorflow.python.keras.callbacks.History at 0x244967948d0>
```

Model Evaluation

At this time, we will evaluate the model on test data.

Code:
```
predictions = model.predict(X_test)
predictions = lb.inverse_transform(predictions)
y_test = lb.inverse_transform(y_test)
from sklearn.metrics import confusion_matrix,classification_report
print(confusion_matrix(y_test,predictions))
print('\n')
print(classification_report(y_test,predictions))
```

Output:
```
[[415    4   11   11]
 [ 13  141   11    2]
 [ 14    2  263    4]
 [ 20    1    5  168]]
```

	precision	recall	f1-score	support
Bacterial_Infections_and_Mycoses	0.90	0.94	0.92	441
Eye_Diseases	0.95	0.84	0.90	167
Skin_and_Connective_Tissue_Diseases	0.91	0.93	0.92	283
Virus_Diseases	0.91	0.87	0.89	194
accuracy			0.91	1085
macro avg	0.92	0.90	0.90	1085
weighted avg	0.91	0.91	0.91	1085

The model achieved 91% accuracy on the test set, which is 35% better than the best of the two models above.

New Prediction

For prediction, we will use an abstract published recently in July 2020 which describes the effect of the coronavirus disease. (Bert et al., 2020)

Code:
```
text = 'Memory T cells induced by previous pathogens can shape susceptibility
to, and the clinical severity of, subsequent infections. Little is known about
the presence in humans of pre-existing memory T cells that have the potential
```

```
to recognize severe acute respiratory syndrome 2 (SARS-CoV-2). Here we studied
T cell responses against the structural (nucleocapsid (N) protein) and n
on-structural (NSP7 and NSP13 of ORF1) regions of SARS-CoV-2'
text = text_processing(text)
text = tfidf_transformer.transform(cv.transform([text]))
text.toarray()
text.shape
```

Output:
```
(1, 25953)
```

First, we will transform this text using the TF-IDF transformer and CountVectorizer - which were previously used for fitting and transforming our dataset. Second, the array will be passed to the "predict" method of the model. Third, the label can be retrieved from the label binarizer object using the "inverse_transform" method.

Code:
```
text = text.toarray()
prediction = model.predict(text)
prediction = lb.inverse_transform(prediction)
print(prediction)
```

Output:
```
['Virus_Diseases']
```

We can observe that the model has predicted the class of the abstract correctly, even though it was trained on a corpus of 1994. We can also determine that the abstract does not even contain the term virus in it. This shows the application of NLP along with the power of deep learning

---

Exercise

1. What is the application of natural language processing?
2. What is the bag-of-words text vectorization method?
3. How does TF-IDF find the significance of a word in the text and the corpus?
4. What is the difference between label encoding and label binarization?

REFERENCES

Bert, N. L., Tan, A. T., Kunasegaran, K., Tham, C. Y. L., Hafezi, M., Chia, A., Chng, M. H. Y., Tan, N., Linster, M., Chia, W. N., Chen, M. I.-C., Wang, L.-F., Ooi, E. E., Kalimuddin, S., Tambyah, P. A., Low, J. G.-H., & Tan, Y.-J. (2020). SARS-CoV-2-specific T-cell immunity in cases of COVID-19 and SARS, and uninfected controls. *Nature*, 584, 457–462. https://doi.org/10.1038/s41586-020-2550-z.

Hersh, W., Buckley, C., Leone, T. J., & Hickam, D. (1994). OHSUMED: An interactive retrieval evaluation and new large test collection for research. *Proceedings of the 17th Annual International ACM SIGIR Conference on Research and Development in Information Retrieval, SIGIR 1994*. https://doi.org/10.1007/978-1-4471-2099-5_20.

16

K-Means Clustering

Introduction

In the Machine Learning section until this chapter, we have discussed supervised learning - where labels of instances were present, and we were tasked to build classifiers using features and labels. On the other hand, in unsupervised learning, the algorithms are not provided with labels. They find the hidden patterns inside the datasets and form clusters by determining similarities within the datasets and by placing objects that are more similar to one another in the same group that is called a cluster. In this chapter, we will discuss one of the popular unsupervised learning algorithms the k-means clustering. K-means clustering follows a simple iterative rule to classify an unlabeled dataset, or more specifically, to form clusters. Clustering has a wide range of applications in biology in studying expression data, drug repurposing, and categorizing organisms or proteins.

In k-means clustering, the first step is to define the value of k, since the algorithm will cluster the data to those many numbers of categories. Selecting the value of k requires domain knowledge, although there are methods, as also discussed in chapter 10, for determining the value of k. We will study these methods for clustering - or unsupervised learning - which we will also discuss in the implementation part of this algorithm.

After deciding the value of k, these are placed randomly inside the dataset space and are called centroids. Data points that are nearer to these randomly placed centroids are assigned to a cluster, such that the sum of the squared distance between the data points and the cluster's centroid (i.e. the arithmetic mean of all of the data points that belong to that cluster) is the minimum.

After assigning the points nearer to the k clusters, the third step is when the centroids are moved towards the center of the cluster, as illustrated in Figure 16.1. The center of a cluster is the mean of every data point belonging to the cluster. After shifting the centroids to the mean of the cluster, the nearer data points are then assigned to k clusters, and the aforementioned steps continue repeating until the centroids stop moving or until the movement is insignificant.

To conclude, we attained well-defined k clusters and observed similar data points assigned to the clusters to discover hidden structures and to derive meaningful insights from the dataset.

The center of the clusters is found by calculating the means of the coordinates or features of the data points. An example is shown in Figure 16.2.

Implementation of K-Means Clustering Using Sklearn

K-means is an unsupervised algorithm, so the dataset provided to k-means will have no labels. In the first analysis, we will generate a dummy dataset with two features and five classes. Next, we will pass the data to the k-means algorithm without the information of classes, and then we will compare the generated clusters to the original cluster of the dummy dataset.

To generate cluster data, sklearn has a library labeled "datasets". This library has a method called "make_blobs". This can create random datasets with a user-defined number of features and clusters.

Code:

```
from sklearn.datasets import make_blobs
# Create Data
```

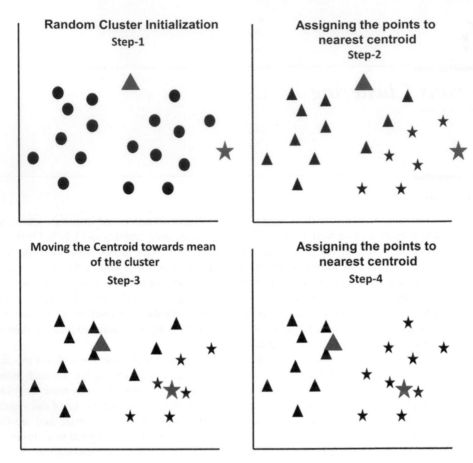

FIGURE 16.1   Steps of *K*-Means Clustering.

Centroid	Nearest Data Points	New Centroid
(3,4)	(2,3) (4,3)	((2+4+3+1)/4, (3+3+2+4)/4)
	(3,2) (1,4)	(2.5,3)

FIGURE 16.2   Calculating the Center of a Cluster.

```
data = make_blobs(n_samples=200, n_features=2,
                             centers=5, cluster_std=1,random_state=150)
plt.scatter(data[0][:,0],data[0][:,1],c=data[1],cmap='rainbow')
plt.xlabel("X1")
plt.ylabel('X2')
```

Output:

The "make_blobs" function acquires arguments like n_samples, n_features, centers, and cluster_std - where "n_samples" are the number of instances; "n_features" are the number of features; "centers" are the number of clusters; "cluster_std" is used to determine the standard deviation within the cluster. In the example, we have selected two features so that we can visualize the clusters. In Figure 16.3, we notice that the five clusters are differently colored.

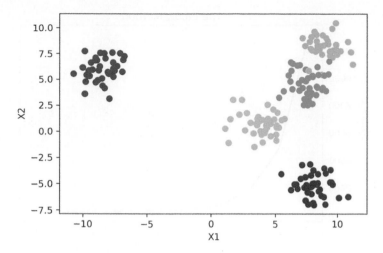

FIGURE 16.3 Randomly Generated Clusters.

Choosing the Number of Clusters

In choosing the optimum value of k, we will train the k-means clustering algorithm with different values of k while observing the within-cluster-sum-of-squares (WCSS). The distance of points to its centroid is calculated and squared. Next, all of the squared distances are summed up to calculate the error.

As the value of K increases, the error decreases. From the WCSS versus k-value plot, we can find the optimum k-value.

Code:
```
from sklearn.cluster import KMeans
wcss = []
for i in range(1, 11):
    kmeans = KMeans(n_clusters = i, init = 'k-means++', random_state = 150)
    kmeans.fit(data[0])
    wcss.append(kmeans.inertia_)
plt.plot(range(1, 11), wcss,color='k')
plt.title('The Elbow Method')
plt.xlabel('Number of clusters')
plt.ylabel('WCSS')
plt.show()
```

Output:
From the elbow plot below (Figure 16.4), it can be observed that the slope of the error remains almost the same after reaching five clusters. Therefore, we will train the model with the k-value of five and analyze the result accordingly.

Code:
```
from sklearn.cluster import KMeans
kmeans = KMeans(n_clusters=5,init = 'k-means++')
kmeans.fit(data[0])
```

Code:
```
f, (ax1, ax2) = plt.subplots(1, 2, sharey=True,figsize=(10,6))
ax1.set_title('K Means')
ax1.scatter(data[0][:,0],data[0][:,1],c=kmeans.labels_,cmap='rainbow')
```

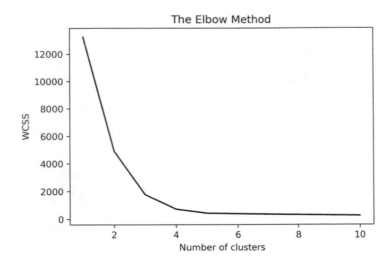

FIGURE 16.4   WCSS versus *K*-Value Plot.

```
ax2.set_title("Original")
ax2.scatter(data[0][:,0],data[0][:,1],c=data[1],cmap='rainbow')
```

Output:

We plotted the results of the clusters formed by *k*-means and the original clusters of data (Figure 16.5a and b). As illustrated, the color difference among the clusters does not signify anything, because they are randomly allotted by the Matplotlib library. However, the area and the structure of the original clusters and *k*-means predicted clusters are mostly similar with minute errors in the overlapping regions.

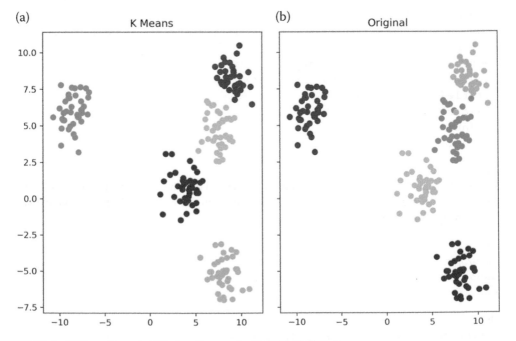

FIGURE 16.5   *K*-Means Generated Clusters Compared with Original Clusters.

TABLE 16.1

Dataframe's Head for DEG Data

	GB_ACC	Control_1	Control_2	Control_3	Control_4	Treated_1	Treated_2	Treated_3	Treated_4
0	NM_080863	8.163398	7.315602	7.305606	8.657140	9.642052	8.896332	9.869440	8.719047
1	BU682208	7.779391	7.172927	3.944858	6.507795	7.837943	8.327777	8.751544	7.235536
2	BC012528	8.587215	7.062856	6.277985	4.336283	8.104861	8.756223	9.707532	7.955940
3	BC036407	8.820179	9.316734	9.170426	7.210428	9.704077	10.421329	11.122569	9.643676
4	BC029869	7.830990	4.209453	6.137504	4.078951	7.001127	7.326429	8.431289	7.835419

We can retrieve the centers of clusters using the attribute "cluster_centers_" of the "KMeans" object.

Code:
```
kmeans.cluster_centers_
```

Output:
```
array([[ 3.89242402, 0.72606842 ],
       [-8.53502186, 5.98022091 ],
       [ 7.51724497, 4.61232782 ],
       [ 8.1037214, -5.1507853 ],
       [ 9.01735573, 8.23432609]])
```

K-Means Clustering of Genes Based on the Co-Expression

Unsupervised clustering has several applications in gene expression analysis. Genes co-expressed under certain conditions are assumed to have some associations. At this point, we will use the differentially expressed genes, which we determined while analyzing microarray data in chapter 7.

Code:
```
dataset = pd.read_csv('DEGs.csv')
dataset.head()
```

Output:
The DEG dataset contains the column as GenBank accession numbers, and the rest of the eight columns present the expression values of the four controls and the four treated samples (Table 16.1).

Code:
```
dataset.describe()
```

Output:
There are 132 genes, which are to be clustered based on their expression under a particular condition (i.e. control-treated) (Table 16.2).

Code:
```
dataset.isna().sum()
```

Output:
```
 GB_ACC        0
Control_1      0
Control_2      0
Control_3      0
```

TABLE 16.2

The Output of Described Method Applied on DEG Data

	Count	Mean	Std	Min	25%	50%	75%	Max
Control_1	132.0	9.070583	1.278072	6.685099	8.120952	8.827167	9.985888	12.353533
Control_2	132.0	8.508938	1.767599	3.277985	7.395747	8.312169	9.699045	12.883617
Control_3	132.0	8.248652	2.001190	1.722466	7.269044	8.144647	9.618288	12.502359
Control_4	132.0	8.055377	2.197749	2.485427	6.651550	8.276980	9.548193	12.716798
Treated_1	132.0	8.402029	2.200985	1.887525	7.097323	8.735816	9.843254	12.905443
Treated_2	132.0	8.117435	2.345779	2.201634	6.638132	8.513519	9.802632	13.301425
Treated_3	132.0	8.295692	2.321438	2.201634	6.978917	8.680008	9.870595	13.024049
Treated_4	132.0	8.542134	2.100473	2.104337	7.448343	8.907190	9.726872	12.892391

```
Control_4     0
Treated_1     0
Treated_2     0
Treated_3     0
Treated_4     0
dtype: int64
```

The dataset contains no null values. Next, we will separate the expression values for the treated samples. There are four expression values for each gene in the treated condition, hence, four features for each gene (Table 16.3).

Code:
```
new_data=dataset.iloc[:,5:]
new_data.head()
```
Output:
Using the elbow method, we will determine the optimal value of *k*.

Code:
```
from sklearn.cluster import KMeans
wcss = []
for i in range(1, 20):
    kmeans = KMeans(n_clusters = i, init = 'k-means++', random_state = 150)
    kmeans.fit(new_data)
    wcss.append(kmeans.inertia_)
plt.plot(range(1, 20), wcss,color='k')
plt.title('The Elbow Method')
```

TABLE 16.3

Expression Values of Four Treated Samples

	Treated_1	Treated_2	Treated_3	Treated_4
0	9.642052	8.896332	9.869440	8.719047
1	7.837943	8.327777	8.751544	7.235536
2	8.104861	8.756223	9.707532	7.955940
3	9.704077	10.421329	11.122569	9.643676
4	7.001127	7.326429	8.431289	7.835419

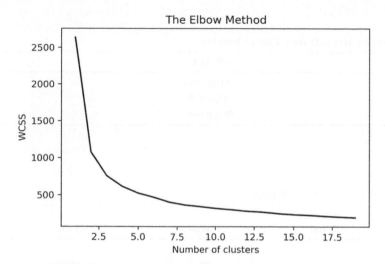

FIGURE 16.6 Elbow Method for Determining the Number of *K*s for Gene Data Clustering.

```
plt.xlabel('Number of clusters')
plt.ylabel('WCSS')
plt.show()
```

Output:
From the elbow method, the number of optimal clusters is likely to be eight (Figure 16.6).

Code:

```
kmeans.labels_
```

Output:
```
array([7, 7, 7, 1, 2, 1, 5, 1, 5, 2, 0, 7, 7, 1, 1, 7, 1, 1, 1, 7, 1, 0,
       7, 5, 5, 1, 1, 7, 7, 1, 1, 4, 1, 7, 2, 2, 7, 7, 7, 5, 1, 5, 2, 2,
       2, 1, 1, 2, 7, 5, 2, 7, 1, 1, 1, 1, 7, 2, 1, 7, 7, 1, 5, 1, 7, 2,
       2, 4, 5, 6, 1, 4, 4, 0, 0, 6, 6, 4, 3, 7, 3, 4, 0, 1, 2, 0, 3, 7,
       2, 2, 3, 0, 1, 2, 6, 6, 3, 7, 0, 2, 1, 2, 1, 3, 7, 7, 0, 4, 1, 4,
       7, 0, 2, 6, 3, 7, 7, 2, 2, 0, 0, 0, 1, 4, 0, 2, 4, 6, 4, 6, 6, 6])
```

In the next step, we will assign the cluster numbers retrieved from the step above to the GenBank accession IDs.

Code:
```
cluster_genes =
pd.concat([dataset.iloc[:,0],pd.DataFrame(kmeans.labels_)],axis=1)
cluster_genes.columns= ['GB_ACC','Groups']
cluster_genes.head()
```

Output:
Thereafter, the genes are grouped into respective clusters using the Pandas dataframe's "groupby function", which takes the column name on the basis of where the user wants to cluster the data (Table 16.4).

Code:
```
cluster = cluster_genes.groupby('Groups')
cluster.describe()
```

TABLE 16.4

GeneBank Accession IDs with their Cluster Number

	GB_ACC	Groups
0	NM_080863	7
2	BC012528	7
4	BC029869	2

TABLE 16.5

Count of Genes in Each Cluster or Group

	GB_ACC Count	Unique	Top	Freq.
Groups				
0	14	14	AI192838	1
1	31	31	T16257	1
2	22	22	NM_016831	1
3	7	7	BF732767	1
4	11	11	AF397394	1
5	9	9	AI580142	1
6	10	10	H79994	1
7	28	28	W03928	1

In Table 16.5, we can observe that Cluster 1 has the highest number of genes (i.e. 31). The code below can be used to retrieve the genes in a particular group. For example, we retrieve all of the genes of Cluster 3 below:

Code:

```
cluster = {k: v for k, v in cluster_genes.groupby('Groups')}
cluster[3]
```

Output:

Corroboration with the literature and the pathway analysis of these genes, along with the gene enrichment analysis can aid in finding the actual association of these genes and derive meaningful biological insights (Table 16.6).

TABLE 16.6

Genes Assigned to Cluster 3

	GB_ACC	Groups
78	BC028378	3
80	BC035640	3
86	NM_006486	3
90	BC005248	3
96	NM_002301	3
103	NM_016619	3
114	BF732767	3

Exercise

1. Explain the algorithm behind the *k*-means clustering.
2. How is the number of clusters chosen for *k*-means clustering on unlabeled data?
3. Generate dummy data of eight clusters and apply *k*-means clustering on it. Next, compare the original clusters with the *k*-means predicted cluster.
4. Use the practice gene expression data from chapter 7, and try means clustering to find the co-expressed genes for SARS infection.
5. Perform gene enrichment analysis for clusters individually to find biological correlations among them.

Index

Printed and bound by CPI Group (UK) Ltd, Croydon, CR0 4YY

24/10/2024

01778288-0009